Springer Tracts in Modern Physics

Volume 249

For further volumes:
http://www.springer.com/series/426

Springer Tracts in Modern Physics

Springer Tracts in Modern Physics provides comprehensive and critical reviews of topics of current interest in physics. The following fields are emphasized: elementary particle physics, solid-state physics, complex systems, and fundamental astrophysics.

Suitable reviews of other fields can also be accepted. The editors encourage prospective authors to correspond with them in advance of submitting an article. For reviews of topics belonging to the above mentioned fields, they should address the responsible editor, otherwise the managing editor.

See also springer.com

Preface

The Tevatron has been the collider with the highest center of mass energy for a long time. The running at the energy of approximately 2 TeV allowed to create heavy particles, like the top quark whose discovery was one of the main Tevatron results. But the Tevatron experiments CDF and D0 also made outstanding contributions on the flavor physics sector by exploiting the huge samples of bottom and charm hadrons produced in the $p\bar{p}$ collisions.

Studies of physics processes involving a change of a flavor quantum number have often lead to new insights. Our current knowledge about bottom and charm hadrons was mainly obtained at e^+e^- colliders and hadron colliders, like the Tevatron and the LHC. The two approaches are complementary in the sense that some measurements can only be done at one type of collider experiment and not or hardly at the other. The different experimental conditions also require different techniques in the detector construction and the data analysis.

This book gives an overview of the flavor physics results of the Tevatron experiments CDF and D0 and the employed experimental techniques. The results published until mid of the year 2012 are covered. A few further results will become available after this date, mainly analyses that are updated to the full Tevatron dataset, and also some new ones. However, no significantly new insights are expected from them. While several flavor physics measurements will be or are already dominated by LHCb, the underlying physics processes are basically the same as at the Tevatron. Hence, also the measurement techniques described here are often reused at the LHC.

Although some discussion of the theoretical background is provided in this book, it focuses on the experimental results and measurement techniques. Knowledge of quantum field theory is not required, but it is assumed that the reader is familiar with the basic concepts of particle physics.

After a brief introduction in Chap. 1, the description of flavor in the standard model is discussed in Chap. 2. The formulas to describe mixing and *CP* violation phenomena are derived in Chap. 3. The Tevatron accelerator and the CDF and D0 detectors are presented in Chap. 4. The following Chaps. 5–9 cover measurements of lifetimes and decay branching fractions, oscillations, indirect *CP* violation, direct *CP* violation, and rare decays. A conclusion is given in the last Chapter.

I would like to thank the persons who have supported me and made it possible to produce this book, Michael Feindt, Thomas Müller, Ulrich Nierste, my colleagues in the CDF B group, and Michal Kreps with whom I worked together on several measurements that entered this book. In particular, I thank my wife, Jeannine, for sharing the good and bad times in particle physics and life.

Contents

Chapter 1
Introduction

Flavor has proven a very successful concept for the description of elementary particles and their interaction. Since the early days of particle physics unexpected experimental results on the flavor physics sector have triggered progress in our theoretical understanding and theoretical predictions on this field have inspired experimental programs.

One of these surprising results was the observation of so called "strange" particles with lifetimes much higher than expected for strong or electromagnetic decays. In 1953 Gell-Mann introduced the flavor quantum number *strangeness* to solve this mystery [1]. He postulated that the strangeness is conserved in strong and electromagnetic interactions and can only be changed in weak interactions. This not only explained the long lifetimes of strange particles, but also the fact that strange particles are always produced in pairs in strong reactions.

Furthermore the strangeness quantum number allowed to establish a systematic characterization of the many different kinds of particles discovered so far. Together with the isospin, a quantum number introduced to describe the symmetry between protons and neutrons, two-dimensional multiplets of particle states were constructed. However not all places in these multiplets could be matched to particles known at that time. The discovery of the Ω^- [2] in 1964 filled the last empty place and confirmed Gell-Mann's idea in an impressive way. He was awarded the Nobel Prize five years later.

In 1964 Gell-Mann and Zweig postulated the quarks as elementary building blocks of hadrons and carriers of the flavor quantum numbers [3, 4]. At that time three types of quarks were known, the up-quark with electric charge $+2/3$ and the down- and strange-quarks with charge $-1/3$. Based on this set of quarks a branching ratio of neutral kaons to muon pairs was calculated that was orders of magnitude higher than the measured value. This discrepancy was solved by Glashow, Iliopoulos, and Maiani in 1970 by introducing a fourth quark flavor, the charm quark [5]. The destructive interference of the processes with up- and charm-quarks brought the prediction in good agreement with the experiment. In 1974 Richter and Ting presented the experimental evidence that the charm quark was not just a theoretical idea, but a real

T. Kuhr, *Flavor Physics at the Tevatron*, Springer Tracts in Modern Physics 249,
DOI: 10.1007/978-3-642-10300-1_1,

particle [6, 7]. For the discovery of the J/ψ particle they were awarded the Nobel Prize two years later.

Despite the progress in the understanding of the subatomic world by the introduction of the charm quark there were still unexplained experimental results. In 1964 Cronin and Fitch had observed the decay of long lived neutral kaons to two pions [8]. While the K_L^0 was believed to be an eigenstate of the combined parity and charge-conjugation operation (CP) with negative eigenvalue the two pions are a CP even eigenstate. The observed decay therefore showed that the CP-parity is not conserved in processes of the weak interaction. For the discovery of this small effect of the order of 10^{-3} Cronin and Fitch were awarded the Nobel Prize in 1980.

Again an extension of the set of quark flavors was able to bring this experimental result in agreement with theory. In 1972, two years before the charm quark was discovered, Kobayashi and Maskawa introduced two more quarks [9]. The bottom quark was discovered 1977 by the E288 collaboration [10] and the top quark 1995 by the CDF and D0 collaborations [11, 12].

In addition to the prediction of these quarks Kobayashi and Maskawa introduced one parameter in the model that could explain the observed CP-violation. Now it was the task of the experimental physicists to search for CP-violation in other processes than the K_L^0 decay and check whether it was consistent with the so-called KM-mechanism. It took almost thirty years until in 2001 the BaBar and Belle collaborations announced evidence for CP-violation in the B^0 system [13, 14] and confirmed the prediction of Kobayashi and Maskawa. For this success of their flavor theory they were awarded the Nobel Prize in 2008.

This theory which describes the relation between quark flavors in a 3×3 matrix named after Cabibbo-Kobayashi-Maskawa is still state of the art and an integral part of the standard model of particle physics. The CKM matrix has four free parameters that are not predicted by any theory, but have to be measured experimentally. The precise determination of these fundamental parameters of nature is one goal of the flavor physics experimental program. The other important task is to check measurements for their consistency with theory. If evidence for a discrepancy between an experimental result and the prediction of the theory of Kobayashi and Maskawa with parameters constrained by other measurements could be established this would be a clear sign of physics beyond the standard model.

This search for new physics is inspired by cosmological considerations. It is assumed that in the big bang an equal amount of matter and anti-matter was produced. The fact that today we have basically only matter left in the universe requires processes that are asymmetric under CP transformation as shown by Sakharov in 1967 [15]. CP violation is therefore an essential condition for our existence. However the CP violation in the standard model is not sufficient by many orders of magnitude to account for the matter antimatter asymmetry in the universe [16]. For this reason measurements of CP violation are promising candidates to find new physics and foster our understanding of nature.

Different experimental approaches are followed to study flavor physics. The so-called B-factories produce pairs of $B^0/\bar{B}^0$ and B^+/B^- mesons in e^+e^- collisions at the $\Upsilon(4S)$ center of mass energy. While these reactions provide well-defined kinematic conditions and have low background, they do not give access to heavier particles. Such particles, like B_s^0 mesons and baryons containing bottom quarks, are copiously produced in $p\bar{p}$ collisions at the Tevatron. Although the signal extraction is more challenging than at B-factories this opens unique opportunities for measurements such as the *CP*-violation in the B_s^0 system and the search for new physics in $B_s^0 \to \mu^+\mu^-$ decays.

This book gives a review of the flavor physics results obtained at the upgraded Tevatron accelerator (Run-II) up to now, a time the LHCb experiment is taking over the lead in flavor physics at a hadron collider. The experimental techniques developed by CDF and D0 are the basis for many of the recent flavor physics results at the LHC.

References

1. M. Gell-Mann, Phys. Rev. **92**, 833 (1953)
2. V.E. Barnes et al., Phys. Rev. Lett. **12**, 204 (1964)
3. M. Gell-Mann, Phys. Lett. **8**, 214 (1964)
4. G.Zweig, CERN-TH-412 (1964)
5. S.L. Glashow, J. Iliopoulos, L. Maiani, Phys. Rev. **D2**, 1285 (1970)
6. J.E. Augustin et al., Phys. Rev. Lett. **33**, 1406 (1974)
7. J.J. Aubert et al., Phys. Rev. Lett. **33**, 1404 (1974)
8. J.H. Christenson, J.W. Cronin, V.L. Fitch, R. Turlay, Phys. Rev. Lett. **13**, 138 (1964)
9. M. Kobayashi, T. Maskawa, Prog. Theor. Phys. **49**, 652 (1973)
10. S.W. Herb et al., Phys. Rev. Lett. **39**, 252 (1977)
11. F. Abe et al., Phys. Rev. Lett. **74**, 2626 (1995)
12. S. Abachi et al., Phys. Rev. Lett. **74**, 2632 (1995)
13. B. Aubert et al., Phys. Rev. Lett. **87**, 091801 (2001)
14. K. Abe et al., Phys. Rev. Lett. **87**, 091802 (2001)
15. A.D. Sakharov, Pisma. Zh. Eksp. Teor. Fiz. **5**, 32 (1967)
16. M.E. Shaposhnikov, JETP Lett. **44**, 465 (1986)

Chapter 2
Flavor in the Standard Model

2.1 The CKM Matrix

The standard model of particle physics describes the fundamental particles and their interactions, the strong, weak and electromagnetic force, very successfully. An important quantum number in this model is the flavor of quarks, introduced by Gell-Mann in 1953 [1]. The three generations of quark flavor pairs are:

$$\begin{pmatrix} \mathbf{u}\,\mathrm{p} \\ \mathbf{d}\,\mathrm{own} \end{pmatrix} \begin{pmatrix} \mathbf{c}\,\mathrm{harm} \\ \mathbf{s}\,\mathrm{trange} \end{pmatrix} \begin{pmatrix} \mathbf{t}\,\mathrm{op} \\ \mathbf{b}\,\mathrm{ottom} \end{pmatrix} \tag{2.1}$$

Each pair consists of an up-type quark with electric charge $+2/3e$ and a down-type quark with charge $-1/3e$. The generations are distinguished by the different masses, increasing from the first to the third.

In the standard model the flavor quantum number is conserved in strong and electromagnetic interactions. It can only be changed by charged current weak processes, described by the exchange of a $W^{\pm}$ boson. The neutral current weak interaction (Z^0 boson exchange) is flavor-conserving. Therefore flavor changing neutral currents (FCNC) do not occur in the standard model at tree level. This makes FCNC processes a good candidate to search for deviations from the standard model because new particles or new interactions may introduce flavor changing tree level amplitudes, that are of comparable size or larger than the amplitude of the higher order standard model loop processes.

The W boson couples to good approximation to the pairs of left handed quarks within one generation as shown in Eq. (2.1). The β decay of the neutron is one example for such a process. But as the decay $K^- \to \pi^0 \mu^- \bar{\nu}_\mu$ demonstrates transitions between generations are possible, too, in this case from a s quark to a u quark. The coupling of the W boson to a us pair, however, is much smaller than to a ud pair. To describe this effect Cabibbo introduced a mixing angle θ_c [2] in order to preserve a common coupling parameter g that is multiplied by the factor $\cos\theta_c$ for ud pairs and

T. Kuhr, *Flavor Physics at the Tevatron*, Springer Tracts in Modern Physics 249,
DOI: 10.1007/978-3-642-10300-1_2,

$\sin\theta_c$ for us pairs. This leads to a 2×2 mixing matrix for the first two generations of quarks.

This model describes the experimental results for processes involving the first two generations already quite well, but it turned out to be only an approximation of a more general theory. The decays $B^0 \to D^-\pi^+$ and $B^0 \to \pi^+\pi^-$ show that there are also transitions from the third to the second and first generation, respectively. The coupling of the W to a cb or ub pair, however, is even more suppressed than the us coupling. To describe the different coupling strengths across the three generations the 2×2 Cabibbo matrix is extended to a 3×3 matrix. The Cabibbo-Kobayashi-Maskawa matrix, or short CKM matrix,

$$V_{CKM} = \begin{pmatrix} V_{ud} & V_{us} & V_{ub} \\ V_{cd} & V_{cs} & V_{cb} \\ V_{td} & V_{ts} & V_{tb} \end{pmatrix} \tag{2.2}$$

determines the coupling of W bosons to pairs of up- (u, c, t) and down-type quarks (d, s, b). The dynamics of the charged current interaction between left handed quarks can be expressed by the Lagrangian

$$\mathscr{L} = -\frac{g}{2}\left(\bar{u}\ \bar{c}\ \bar{t}\right)_L \gamma^\mu V_{CKM} \begin{pmatrix} d \\ s \\ b \end{pmatrix}_L W^+_\mu + \text{h.c.}, \tag{2.3}$$

where the subscript L denotes the left-handed component of the quark fields, γ^μ are the gamma matrices, W^+_μ is the gauge field, and h.c. stands for the hermitian conjugated of the preceding expression.

The basis of quark states can be changed from the mass eigenstates to a basis where the CKM matrix vanishes. Usually the down-type quarks are transformed:

$$\begin{pmatrix} d' \\ s' \\ b' \end{pmatrix} = V_{CKM} \begin{pmatrix} d \\ s \\ b \end{pmatrix}. \tag{2.4}$$

The coupling of the W boson to quarks can then be described by the coupling to ud', cs', and tb' pairs with coupling constant g. The states d', s', and b' are therefore called weak eigenstates, but one should keep in mind that there is an infinite number of weak bases, related by arbitrary unitary transformations. E.g. one could change the basis of up-type quarks instead of the down-type quark basis by

$$\left(\bar{u}'\ \bar{c}'\ \bar{t}'\right) = \left(\bar{u}\ \bar{c}\ \bar{t}\right) V_{CKM} \tag{2.5}$$

and then describe the charged current interaction by the coupling of W bosons to $u'd$, $c's$, and $t'b$ pairs.

The CKM matrix has 9 complex elements. The unitarity condition

$$V_{CKM} V_{CKM}^{\dagger} = V_{CKM}^{\dagger} V_{CKM} = 1 \tag{2.6}$$

leads to nine independent equations and therefore removes nine of the originally 18 free parameters. Furthermore five phases can be eliminated by absorbing them in the quark fields. This leaves four free parameters. These parameters can be chosen as three mixing angles θ_{12}, θ_{23}, and θ_{13} between the three generations and a phase factor δ:

$$\begin{aligned} V_{CKM} &= \begin{pmatrix} 1 & 0 & 0 \\ 0 & c_{23} & s_{23} \\ 0 & -s_{23} & c_{23} \end{pmatrix} \cdot \begin{pmatrix} c_{13} & 0 & s_{13}\, e^{-i\delta} \\ 0 & 1 & 0 \\ -s_{13}\, e^{i\delta} & 0 & c_{13} \end{pmatrix} \cdot \begin{pmatrix} c_{12} & s_{12} & 0 \\ -s_{12} & c_{12} & 0 \\ 0 & 0 & 1 \end{pmatrix} \\ &= \begin{pmatrix} c_{12}c_{13} & s_{12}c_{13} & s_{13}\, e^{-i\delta} \\ -s_{12}c_{23} - c_{12}s_{23}s_{13}\, e^{i\delta} & c_{12}c_{23} - s_{12}s_{23}s_{13}\, e^{i\delta} & s_{23}c_{13} \\ s_{12}s_{23} - c_{12}c_{23}s_{13}\, e^{i\delta} & -s_{23}c_{12} - s_{12}c_{23}s_{13}\, e^{i\delta} & c_{23}c_{13} \end{pmatrix} \end{aligned} \tag{2.7}$$

where $c_{ij} = \cos\theta_{ij}$ and $s_{ij} = \sin\theta_{ij}$ with $ij \in \{12, 23, 13\}$.

Note that the fact which elements are real and which are complex depends on the phase convention. Only phase differences have a physical meaning. The phase differences between certain combinations of CKM matrix elements that arises if δ is not equal to zero or π is responsible for *CP* violation in the quark sector. The connection between phases and experimental observables will be discussed in Chap. 3. The phase δ is not present in the quark mixing matrix for two generations. The achievement of Kobayashi and Maskawa was to show that this one physical phase occurs if a further generation is added. So far all experimentally observed effects of *CP*-violation could be explained by this single parameter.

However, the phase and the mixing angles are not predicted by theory. They have to be measured by experiments. The precise determination of these fundamental parameters of nature is one of the main goals of flavor physics research.

The measurements reveal a hierarchy of quark mixings between generations: $1 \gg \theta_{12} \gg \theta_{23} \gg \theta_{13}$. This structure can be clearly seen if the CKM matrix is written in the Wolfenstein parametrization [3]

$$V_{CKM} = \begin{pmatrix} 1-\lambda^2/2 & \lambda & A\lambda^3(\rho - i\eta) \\ -\lambda & 1-\lambda^2/2 & A\lambda^2 \\ A\lambda^3(1-\rho-i\eta) & -A\lambda^2 & 1 \end{pmatrix} + O(\lambda^4) \tag{2.8}$$

where A, ρ and η are real parameters of order unity and $\lambda := |V_{us}| \approx 0.22$. It can be seen that the diagonal elements are of order one and the off-diagonal elements are of order λ, λ^2, and λ^3 for transitions between first and second, second and third, and first and third generation, respectively. Furthermore one can see that the phase of all elements is very similar (approximately zero in this parametrization), except for V_{ub} and V_{td}, that are involved in transitions between first and third generation and are suppressed by λ^3. This makes hadrons containing b quarks a good place to look for large CP-violating effects.

Why does the CKM matrix have this structure? Well, we do not know. The research on flavor physics still has to answer this question. Maybe future experiments will reveal hints towards a more general theory that can explain the CKM matrix.

2.2 Unitarity Triangles

In principle four measurements are sufficient to determine the four free parameters of the Kobayashi-Maskawa theory. But since there are many more measurements possible this opens a great opportunity to search for physics beyond the standard model. This search for new physics by overconstraining the CKM parameters is one of the main tasks of flavor physics experiments.

If the standard model is valid all measurements must be described by a unitary CKM matrix. In particular Eq. (2.6) implies that the product of different columns vanishes:

$$V_{ud}V_{us}^* + V_{cd}V_{cs}^* + V_{td}V_{ts}^* = 0 \tag{2.9}$$

$$V_{ud}V_{ub}^* + V_{cd}V_{cb}^* + V_{td}V_{tb}^* = 0 \tag{2.10}$$

$$V_{us}V_{ub}^* + V_{cs}V_{cb}^* + V_{ts}V_{tb}^* = 0 \tag{2.11}$$

Analog conditions can be obtained from the product of different rows of the CKM matrix. These relations can be represented by triangles in the complex plane and are known as unitarity triangles.

The triangle obtained from the multiplication of the first and third column, corresponding to Eq. (2.10), is shown in Fig. 2.1. Because all three sides have similar length and all angles are well different from zero it gives a visually better perceivable representation of the unitarity condition than the nearly degenerate triangles corresponding to Eqs. (2.9) and (2.11). It is therefore usually referred to as *the* unitarity triangle. As one can see in Fig. 2.1 the angles in this triangle are

$$\alpha = \arg\left(-\frac{V_{td}V_{tb}^*}{V_{ud}V_{ub}^*}\right), \quad \beta = \arg\left(-\frac{V_{cd}V_{cb}^*}{V_{td}V_{tb}^*}\right), \quad \gamma = \arg\left(-\frac{V_{ud}V_{ub}^*}{V_{cd}V_{cb}^*}\right). \tag{2.12}$$

In order to simplify the display of experimental constraints on the sides and angles of the triangle it is often normalized to $V_{cd}V_{cb}^*$ so that the lower side is fixed to the real axis in the range [0,1] (see Fig. 2.2). After this transformation the coordinates of the apex are given by

$$\bar{\rho} + i\bar{\eta} = -\frac{V_{ud}V_{ub}^*}{V_{cd}V_{cb}^*}. \tag{2.13}$$

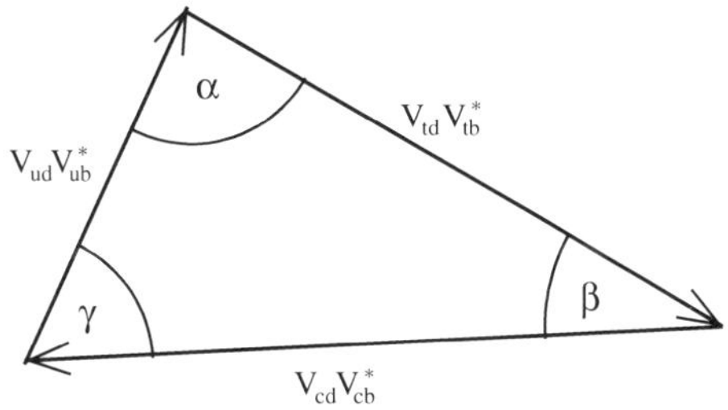

Fig. 2.1 Unitarity triangle for the product of first and third column of the CKM matrix

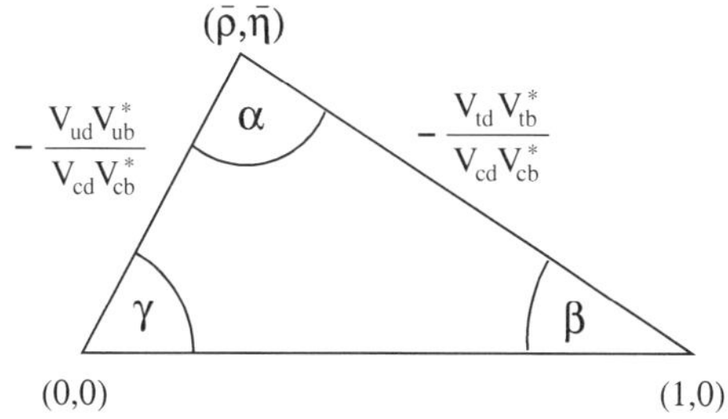

Fig. 2.2 Unitarity triangle for the product of first and third column of the CKM matrix normalized to $V_{cd}V_{cb}^*$

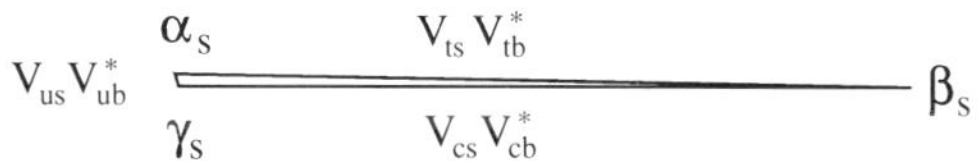

Fig. 2.3 Unitarity triangle for the product of second and third column of the CKM matrix

Since this triangle is constructed from CKM matrix elements for d and b quarks, $b \to d$ transitions, as occurring in the mixing of B^0 and $\bar{B}^0$ mesons which are composed of these two quarks, are ideal to study it. The large angles, corresponding to large phase differences between the involved CKM matrix elements, illustrate that large *CP* violating effects can be expected in the B^0-$\bar{B}^0$system. The situation is different for the triangle obtained from the product of the second and third column, quoted in Eq. (2.11). As can be seen in Fig. 2.3 this triangle, involving the CKM matrix elements for s and b quarks, is nearly degenerate. It gives a visual representation of the small expected *CP* violation in B_s^0-$\bar{B}_s^0$ mixing. Quantitatively this is expressed by the small value of the angle β_s defined as

$$\beta_s = \arg\left(-\frac{V_{ts}V_{tb}^*}{V_{cs}V_{cb}^*}\right). \tag{2.14}$$

Although the unitarity triangles have quite different shapes they all have the same area. The area is a phase convention independent measure of *CP*-violation and equal to half of the Jarlskog invariant J [4] defined by

$$\mathrm{Im}[V_{ij}V_{kl}V_{il}^*V_{kj}^*] = J\sum_{m,n}\varepsilon_{ikm}\varepsilon_{jln}, \tag{2.15}$$

where ε_{ikm} is the antisymmetric permutation tensor. E.g. for $i = 1$, $j = 2$, and $k = l = 3$ one obtains

$$J = -\mathrm{Im}[V_{us} V_{tb} V_{ub}^* V_{ts}^*] \approx -\mathrm{Im}[\lambda \cdot 1 \cdot A\lambda^3(\rho + i\eta) \cdot (-A\lambda^2)] = \lambda^6 A^2 \eta. \quad (2.16)$$

In case of CP-violation J is finite, corresponding to a non-zero value of η in the Wolfenstein parametrization and leading to non-degenerate unitarity triangles. The world average value of the Jarlskog invariant is $J = (2.96^{+0.20}_{-0.16}) \times 10^{-5}$ [5].

The unitarity triangles provide a good visual representation of the consistency checks of the standard model. The sides and angles must match to form a closed triangle if the standard model is correct. As often only *the* unitarity triangle for the product of first and third column of the CKM matrix is shown it should be noted that this gives a necessary, but not a sufficient condition for the standard model consistency. Even if *the* unitarity triangle shows nice consistency there can still be significant new physics contributions in processes involving CKM matrix elements of the second column. To cover as many new physics scenarios as possible, the high precision measurements of B^0 and B^+ mesons at the B factories have to be accompanied by measurements of the B_s^0 system at the Tevatron. In particular the angle β_s is a promising quantity because its value is approximately zero in the standard model and any measurement of a significant non-zero value would be an indication for new physics.

2.3 Theoretical Tools

In flavor physics we are interested in the quark level transitions. However, in nature all quarks are confined in hadrons (except for the top quark that decays before it can hadronize). Therefore we cannot observe quark level processes directly in experiments. We can only measure decays of hadrons containing quarks. This means we have to calculate the transition amplitude, $\mathcal{M}(B \to f)$, of a hadron B to a final state f taking into account the interactions of quarks inside hadrons in order to connect the quark level transitions to the observed hadron decay rates.

Different scales are involved in this calculation that allow to apply a factorization approach and certain approximations. In the standard model flavor changing processes are mediated by a W boson. Because the mass of the W is much larger than the mass of quarks inside the hadron the W exchange takes place at a much shorter distance scale, corresponding to a higher energy scale, than the hadronic interaction. Therefore the "hard" process involving a W or other heavy particles, like the top quark or potential new physics particles, can be factorized and treated as a point-like interaction. This approach is used in the operator product expansion (OPE) [6]. In this framework the decay matrix element $\mathcal{M}$ is given by

$$\mathcal{M}(B \to f) \approx -\frac{4G_F}{\sqrt{2}} V \sum_i C_i(\mu) \langle f | O_i(\mu) | B \rangle . \quad (2.17)$$

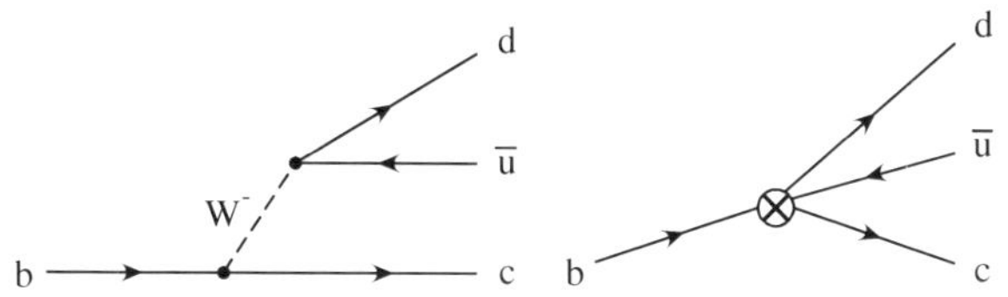

Fig. 2.4 The $b \to c\bar{u}d$ transition described by a W boson exchange (*left*) and by an effective 4-quark operator (*right*)

By convention the constant $4G_F/\sqrt{2}$, where G_F is the Fermi constant, and the involved CKM matrix elements, denoted V, are factored out. The Wilson coefficients $C_i(\mu)$ contain the hard, and the application of the local operators $O_i(\mu)$ to the initial and final states the soft processes. The separation in hard and soft is arbitrary and determined by the introduced scale μ. In principle the sum has an infinite number of terms, but their size is suppressed by powers of the ratio of quark and W masses squared, m_q^2/m_W^2, with increasing dimension. So only a finite number has to be considered in order to get a sufficiently accurate result.

Figure 2.4 illustrates the approximation of the W exchange in a $b \to c\bar{u}d$ transition (left diagram) by a point-like interaction (right diagram). For this tree level process the local operators are given by

$$O_1 = \bar{d}^\alpha \gamma_\mu (1 - \gamma_5) u^\beta \bar{c}^\beta \gamma_\mu (1 - \gamma_5) b^\alpha, \tag{2.18}$$

$$O_2 = \bar{d}^\alpha \gamma_\mu (1 - \gamma_5) u^\alpha \bar{c}^\beta \gamma_\mu (1 - \gamma_5) b^\beta \tag{2.19}$$

where α and β are color indices. The local operators basically provide a projection of the initial and final state hadrons to the quark states, that enter the hard interaction. The operator O_2 describe the case where b and c quarks have the same color charge. The corresponding hard process with Wilson coefficient C_2 is calculated in leading order from the Feynman diagram in Fig. 2.4. The color-rearranged operator O_1 stems from a diagram with an additional gluon connecting the two quark lines.

An advantage of the OPE is that the Wilson coefficients are universal, meaning they are independent of the hadronic final state f. Once they are calculated or measured in one decay mode they can be applied in the prediction for other decay modes. Often the Wilson coefficients are considered effective coupling constants for the point-like interactions described by the local operators. Thus one can use them to define an effective Hamiltonian.

Because of the high scale the Wilson coefficients can be calculated with perturbative methods and are in general well controlled. However, the calculation of the low energetic part is more challenging as non-perturbative methods have to be used.

Often the entire hadronic part is factored out and parametrized by a decay constant or a kinematics dependent form factor. For example the rate of the semileptonic decay $\bar{B}^0 \to D^+\ell^-\bar{\nu}_\ell$ is given by the CKM matrix element V_{cb} from the coupling of the

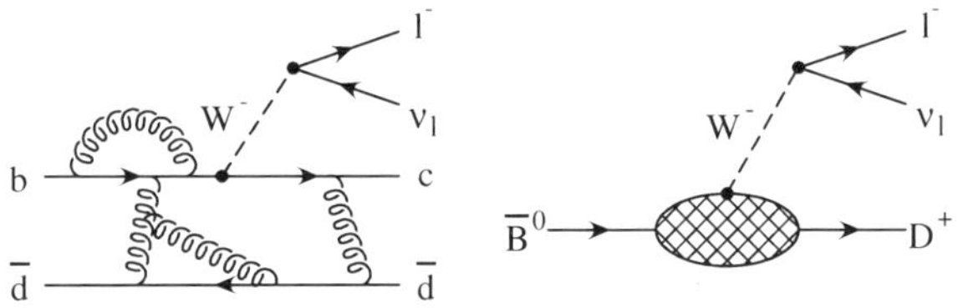

Fig. 2.5 The decay $\bar{B}^0 \rightarrow D^+\ell^-\bar{\nu}_\ell$ at quark level (*left*) and at hadron level described by a form factor (*right*)

W to the b and c quarks and the transition matrix elements between the hadronic systems:

$$\frac{d\Gamma_{\bar{B}^0\rightarrow D^+\ell^-\bar{\nu}_\ell}(q^2)}{dq^2} \sim |V_{cb}|^2 \; |\langle D^+|\bar{c}\gamma^\mu(1-\gamma_5)b|\bar{B}^0\rangle|^2. \tag{2.20}$$

Here q^2 determines the kinematics and is defined by the four-momenta for the $\bar{B}^0$ and D^+ meson as $q^2 = [\mathbf{p}(\bar{B}^0) - \mathbf{p}(D^+)]^2$. When the well-known phase-space factor in the hadronic term in Eq. (2.20) is factored out as $\kappa(q^2)$, the remaining part is described by the form factor $f_{\bar{B}^0\rightarrow D^+}(q^2)$:

$$\frac{d\Gamma_{\bar{B}^0\rightarrow D^+\ell^-\bar{\nu}_\ell}(q^2)}{dq^2} \sim |V_{cb}|^2 \; \kappa(q^2) \; |f_{\bar{B}^0\rightarrow D^+}(q^2)|^2. \tag{2.21}$$

Figure 2.5 illustrates the form factor.

Several approaches exist for the calculation of hadronic matrix elements. The best choice of method depends on the case at hand. A powerful handle to simplify theoretical calculations are symmetries. The QCD Lagrangian is invariant under charge conjugation and parity operation so that the exact symmetries of C and P can be exploited. Strong isospin and $SU(3)$ flavor provide further useful symmetries. As these are only approximately valid there can be small corrections to the ideal case of an exact symmetry.

In the case of bottom and charm quarks, when the mass of the heavy quark, m_Q, is much larger than the QCD scale parameter, Λ_{QCD}, another approximate symmetry arises, the heavy quark symmetry (HQS) [7, 8]. In the limit of an infinite mass, the heavy quark can be treated as a source of a static color field. It becomes independent of the flavor, leading to the heavy quark flavor symmetry. This allows to relate quantities in the charm and bottom sector. Another consequence of an infinite quark mass is the decoupling of the spin, leading to heavy quark spin symmetry.

As the b and even more the c quark have finite mass there are corrections to the HQS limit. They can be calculated in the framework of the Heavy Quark Effective Theory (HQET) [9–11] by an expansion in terms of Λ_{QCD}/m_Q. For the calculation of inclusive decay rates the relation $m_Q \gg \Lambda_{QCD}$ is exploited by the Heavy Quark Expansion (HQE) [12–16].

Further techniques include perturbative QCD methods for exclusive decays, QCD sum rules and models of QCD. Another interesting approach is the numerical calculation of hadronic matrix elements in a discretized space-time, called Lattice QCD [17]. In addition to the progress in the theoretical understanding this method profits from the advances in computing technologies.

References

1. M. Gell-Mann, Phys. Rev. **92**, 833 (1953)
2. N. Cabibbo, Phys. Rev. Lett. **10**, 531 (1963)
3. L. Wolfenstein, Phys. Rev. Lett. **51**, 1945 (1983)
4. C. Jarlskog, Phys. Rev. Lett. **55**, 1039 (1985)
5. J. Beringer et al., Phys. Rev. **D 86**, 010001 (2012)
6. K.G. Wilson, Phys. Rev. **179**, 1499 (1969)
7. N. Isgur, M.B. Wise, Phys. Lett. B**232**, 113 (1989)
8. N. Isgur, M.B. Wise, Phys. Lett. B**237**, 527 (1990)
9. E. Eichten, B.R. Hill, Phys. Lett. B**243**, 427 (1990)
10. H. Georgi, Phys. Lett. B**240**, 447 (1990)
11. B. Grinstein, Nucl. Phys. B**339**, 253 (1990)
12. J. Chay, H. Georgi, B. Grinstein, Phys. Lett. B**247**, 399 (1990)
13. I.I.Y. Bigi, N.G. Uraltsev, A.I. Vainshtein, Phys. Lett. B**293**, 430 (1992)
14. I.I.Y. Bigi, M.A. Shifman, N.G. Uraltsev, A.I. Vainshtein, Phys. Rev. Lett. **71**, 496 (1993)
15. A.V. Manohar, M.B. Wise, Phys. Rev. **D49**, 1310 (1994)
16. B. Blok, L. Koyrakh, M.A. Shifman, A.I. Vainshtein, Phys. Rev. **D49**, 3356 (1994)
17. K.G. Wilson, Phys. Rev. **D10**, 2445 (1974)

Chapter 3
Theoretical Description of Mixing and *CP*-Violation

3.1 Neutral Meson Mixing

One of the most interesting particles to study in flavor physics are neutral mesons that can oscillate into their anti-particle. This effect was first observed for K^0 mesons [1] and later for B^0 [2, 3] and B_s^0 mesons [4]. Recently the oscillation of D^0 mesons was observed, too [5–7].

In this section we want to discuss the quantum mechanical description of the neutral meson system. The flavor eigenstate of the particle that is produced in processes of strong interaction will be denoted $|B\rangle$ and stands for either $|K^0\rangle$, $|D^0\rangle$, $|B^0\rangle$, or $|B_s^0\rangle$. Its anti-particle, $|\bar{B}\rangle$, is obtained by *CP* conjugation:

$$CP|B\rangle = e^{i\xi}|\bar{B}\rangle. \tag{3.1}$$

The transformation introduces a phase ξ, but as this phase is not observable and therefore unphysical we set it to zero without loss of generality. Particle and anti-particle have well defined, opposite flavor quantum numbers and are orthogonal states:

$$\langle B|\bar{B}\rangle = 0. \tag{3.2}$$

A pure flavor eigenstate $|\psi(t=0)\rangle = |B\rangle$ at time $t = 0$ will evolve with time into a mixture of particle and anti-particle and decay to several final states $f_1, f_2, \ldots$:

$$|\psi(t)\rangle = c(t)|B\rangle + \bar{c}(t)|\bar{B}\rangle + c_1(t)|f_1\rangle + c_2(t)|f_2\rangle + \cdots, \tag{3.3}$$

where the coefficients $c(t)$, $\bar{c}(t)$, $c_j(t)$ describe the time-dependent amplitudes of the individual components. If we are only interested in the contributions of the neutral meson states, $c(t)$ and $\bar{c}(t)$, and not in the amplitudes of the decay final states, $c_j(t)$,

T. Kuhr, *Flavor Physics at the Tevatron*, Springer Tracts in Modern Physics 249,
DOI: 10.1007/978-3-642-10300-1_3,

the Wigner-Weisskopf approximation can be used to define a Schrödinger equation with an effective 2×2 matrix:

$$i\frac{d}{dt}\begin{pmatrix} c(t) \\ \bar{c}(t) \end{pmatrix} = \mathscr{H}_{\mathit{eff}} \begin{pmatrix} c(t) \\ \bar{c}(t) \end{pmatrix} \tag{3.4}$$

Since this two-state system neglects the decay final states and thus does not conserve probability the matrix $\mathcal{H}_{eff}$ is non-hermitian. But, as any matrix, it can be expressed in the following way by two hermitian matrices, a mass matrix M and a decay matrix Γ:

$$\mathscr{H}_{\mathit{eff}} = M - \frac{i}{2}\Gamma = \begin{pmatrix} M_{11} & M_{12} \\ M_{12}^{*} & M_{22} \end{pmatrix} - \frac{i}{2}\begin{pmatrix} \Gamma_{11} & \Gamma_{12} \\ \Gamma_{12}^{*} & \Gamma_{22} \end{pmatrix} \tag{3.5}$$

The diagonal elements of M and Γ govern the flavor-conserving transitions while the off-diagonal elements are responsible for flavor-changing transitions.

In the following we will assume that CPT is a fundamental symmetry of nature. This assumption is in good agreement with all experimental results and it is a requirement for a local Lorentz-invariant quantum field theory, a very basic concept we believe describes nature. For meson mixing, CPT invariance implies $M_{11} = M_{22}$ and $\Gamma_{11} = \Gamma_{22}$. Then the eigen-vectors of the effective Hamiltonian are linear combinations of the two flavor eigenstates that are defined by two complex numbers p and q:

$$|B_L\rangle = p|B\rangle + q|\bar{B}\rangle \tag{3.6}$$

$$|B_H\rangle = p|B\rangle - q|\bar{B}\rangle. \tag{3.7}$$

with

$$|p|^2 + |q|^2 = 1. \tag{3.8}$$

The two states $|B_L\rangle$ and $|B_H\rangle$ each have a defined mass and lifetime and are thus usually referred to as mass or lifetime eigenstates. The masses and decay widths are given by the eigenvalues ω_H and ω_L:

$$m_{L/H} = \mathrm{Re}(\omega_{L/H}) \tag{3.9}$$

$$\Gamma_{L/H} = 2\mathrm{Im}(\omega_{L/h}). \tag{3.10}$$

$$\Leftrightarrow \quad \omega_{L/H} = m_{L/H} + i\frac{\Gamma_{L/h}}{2}. \tag{3.11}$$

The two solutions can be distinguished by their different mass or their different lifetime. Here the mass is used as indicated by the indices L and H for light and heavy, respectively. This is the usual nomenclature for B^0 and B_s^0 mesons. In the K^0 system the states have very different decay rates and are therefore labeled by their lifetime, K_L^0 and K_S^0 for the long and short lived state, respectively.

Table 3.1 World average values of $x = \Delta m/\Gamma$ and $y = \Delta\Gamma/2\Gamma$ for weakly decaying neutral mesons [9]

Meson	K^0	D^0	B^0	B_s^0
$x = \Delta m/\Gamma$	0.946 ± 0.002	$0.0063^{+0.0019}_{-0.0020}$	0.770 ± 0.008	26.49 ± 0.29
$y = \Delta\Gamma/2\Gamma$	0.997 ± 0.004	-0.0075 ± 0.0012	0.007 ± 0.009	0.147 ± 0.019

Instead of the mass and width values of the heavy and light states often the mean value and the difference are quoted:

$$m = \frac{m_L + m_H}{2} \qquad \Gamma = \frac{\Gamma_L + \Gamma_H}{2} \tag{3.12}$$

$$\Delta m = m_H - m_L \qquad \Delta\Gamma = \Gamma_L - \Gamma_H. \tag{3.13}$$

The sign of Δm is positive by definition. The sign of $\Delta\Gamma$ has to be measured. Here one has to be careful about the sign convention. Both possibilities, $\Gamma_L - \Gamma_H$ and $\Gamma_H - \Gamma_L$, are used in the literature. For the definition in Eq. (3.13) a positive value of $\Delta\Gamma$ is expected in the standard model for the B^0 and B_s^0 systems.

The mass and decay width differences, as well as the factors p and q defining the heavy and light eigenstates, are determined by the off-diagonal elements of the mass and decay matrices. The following relations are derived in reference [8]:

$$\Delta m^2 - \frac{1}{4}\Delta\Gamma^2 = 4|M_{12}|^2 - |\Gamma_{12}|^2 \tag{3.14}$$

$$\Delta m \Delta\Gamma = -4\mathrm{Re}(M_{12}\Gamma_{12}^*) \tag{3.15}$$

$$\frac{q}{p} = \frac{\Delta m + i\Delta\Gamma/2}{2M_{12} - i\Gamma_{12}} = \frac{2M_{12}^* - i\Gamma_{12}^*}{\Delta m + i\Delta\Gamma/2}. \tag{3.16}$$

If $|\Gamma_{12}| \ll |M_{12}|$ and $\Delta\Gamma \ll \Delta m$, as it is the case in the B^0 and B_s^0 systems, Eqs. (3.14) and (3.15) can be simplified to

$$\Delta m \approx 2|M_{12}| \tag{3.17}$$

$$\Delta\Gamma \approx 2|\Gamma_{12}|\cos\phi \tag{3.18}$$

where ϕ is the phase between the off-diagonal elements of the mass and decay matrix:

$$\phi = \arg\left(-M_{12}\Gamma_{12}^*\right) = \arg\left(\frac{-M_{12}}{\Gamma_{12}}\right). \tag{3.19}$$

As we will see later in this Chapter the phase plays an important role in *CP* violation. The parameters Δm and $\Delta\Gamma$ are essential for the time evolution of the neutral mesons. Table 3.1 gives an overview of the current experimental knowledge of these parameters.

Having determined eigen-values and -vectors of the effective Hamiltonian in Eq. (3.5) the time evolution of these eigenstates is according to Eq. (3.4) described by

$$\frac{d}{dt}|B_{L/H}(t)\rangle = i\omega_{L/H}|B_{L/H}(t)\rangle \tag{3.20}$$

Thus the wave function at time t of an initial heavy or light mass eigenstates is simply given by

$$|B_L(t)\rangle = e^{-i\omega_L t}|B_L\rangle = e^{-i\omega_L t}(p|B\rangle + q|\bar{B}\rangle) \tag{3.21}$$

$$|B_H(t)\rangle = e^{-i\omega_H t}|B_H\rangle = e^{-i\omega_H t}(p|B\rangle - q|\bar{B}\rangle) \tag{3.22}$$

So the mass eigenstates decay exponentially and do not oscillate.

In order to calculate the time evolution of an initial flavor eigenstate we use Eqs. (3.6) and (3.7) to express the flavor by the mass eigenstates:

$$|B\rangle = \frac{1}{2p}(|B_L\rangle + |B_H\rangle) \tag{3.23}$$

$$|\bar{B}\rangle = \frac{1}{2q}(|B_L\rangle - |B_H\rangle). \tag{3.24}$$

Using Eqs. (3.22) and (3.21) we can derive the time-dependent state vector

$$\begin{aligned}|B(t)\rangle &= \frac{1}{2p}\left[|B_L(t)\rangle + |B_H(t)\rangle\right]\\ &= \frac{1}{2p}\left[e^{-i\omega_L t}(p|B\rangle + q|\bar{B}\rangle) + e^{-i\omega_H t}(p|B\rangle - q|\bar{B}\rangle)\right]\\ &= \frac{1}{2}(e^{-i\omega_L t} + e^{-i\omega_H t})|B\rangle + \frac{q}{2p}(e^{-i\omega_L t} - e^{-i\omega_H t})|\bar{B}\rangle\\ &= g_+(t)|B\rangle + \frac{q}{p}g_-(t)|\bar{B}\rangle\end{aligned} \tag{3.25}$$

where

$$g_\pm(t) := \frac{1}{2}\left(e^{-i\omega_L t} \pm e^{-i\omega_H t}\right) = \frac{1}{2}\left(e^{-im_L t - \frac{\Gamma_L}{2}t} \pm e^{-im_H t - \frac{\Gamma_H}{2}t}\right). \tag{3.26}$$

In an analog way one obtains

$$|\bar{B}(t)\rangle = \frac{p}{q}g_-(t)|B\rangle + g_+(t)|\bar{B}\rangle. \tag{3.27}$$

The relations

$$m_{L/H} = m \mp \frac{\Delta m}{2} \quad \text{and} \quad \Gamma_{L/H} = \Gamma \pm \frac{\Delta\Gamma}{2} \tag{3.28}$$

derived from Eqs. (3.12) and (3.13) can be used to express the time dependence in terms of m, Δm, Γ, and $\Delta\Gamma$:

$$\begin{aligned} g_\pm(t) &= \frac{1}{2}\left(e^{-i(m-\frac{\Delta m}{2})t-(\frac{\Gamma}{2}+\frac{\Delta\Gamma}{4})t} \pm e^{-i(m+\frac{\Delta m}{2})t-(\frac{\Gamma}{2}-\frac{\Delta\Gamma}{4})t}\right) \\ &= \frac{1}{2}e^{-imt}e^{-\frac{\Gamma}{2}t}\left(e^{i\frac{\Delta m}{2}t}e^{-\frac{\Delta\Gamma}{4}t} \pm e^{-i\frac{\Delta m}{2}t}e^{\frac{\Delta\Gamma}{4}t}\right) \\ &= \frac{1}{2}e^{-imt}e^{-\frac{\Gamma}{2}t}\left(\cos\frac{\Delta m}{2}t\left[e^{-\frac{\Delta\Gamma}{4}t} \pm e^{\frac{\Delta\Gamma}{4}t}\right] + i\sin\frac{\Delta m}{2}t\left[e^{-\frac{\Delta\Gamma}{4}t} \mp e^{\frac{\Delta\Gamma}{4}t}\right]\right). \end{aligned} \tag{3.29}$$

Expressed by hyperbolic functions this yields [8]:

$$g_+(t) = e^{-imt}e^{-\frac{\Gamma}{2}t}\left(\cosh\frac{\Delta\Gamma}{4}t\cos\frac{\Delta m}{2}t - i\sinh\frac{\Delta\Gamma}{4}t\sin\frac{\Delta m}{2}t\right) \tag{3.30}$$

$$g_-(t) = e^{-imt}e^{-\frac{\Gamma}{2}t}\left(-\sinh\frac{\Delta\Gamma}{4}t\cos\frac{\Delta m}{2}t + i\cosh\frac{\Delta\Gamma}{4}t\sin\frac{\Delta m}{2}t\right). \tag{3.31}$$

For the calculation of decay rates which can be measured experimentally we need the terms $|g_\pm(t)|^2$ and $g_+^*(t)g_-(t)$. They are derived using Eq. (3.29):

$$\begin{aligned} |g_\pm(t)|^2 &= g_\pm^*(t)g_\pm(t) \\ &= \frac{1}{4}e^{-\Gamma t}\left(e^{-i\frac{\Delta m}{2}t}e^{-\frac{\Delta\Gamma}{4}t} \pm e^{i\frac{\Delta m}{2}t}e^{\frac{\Delta\Gamma}{4}t}\right)\left(e^{i\frac{\Delta m}{2}t}e^{-\frac{\Delta\Gamma}{4}t} \pm e^{-i\frac{\Delta m}{2}t}e^{\frac{\Delta\Gamma}{4}t}\right) \\ &= \frac{1}{4}e^{-\Gamma t}\left(e^{-\frac{\Delta\Gamma}{2}t} \pm e^{-i\Delta m t} \pm e^{i\Delta m t} + e^{\frac{\Delta\Gamma}{2}t}\right) \\ &= \frac{1}{2}e^{-\Gamma t}\left(\cosh\frac{\Delta\Gamma}{2}t \pm \cos\Delta m t\right) \end{aligned} \tag{3.32}$$

$$\begin{aligned} g_+^*(t)g_-(t) &= \frac{1}{4}e^{-\Gamma t}\left(e^{-i\frac{\Delta m}{2}t}e^{-\frac{\Delta\Gamma}{4}t} + e^{i\frac{\Delta m}{2}t}e^{\frac{\Delta\Gamma}{4}t}\right)\left(e^{i\frac{\Delta m}{2}t}e^{-\frac{\Delta\Gamma}{4}t} - e^{-i\frac{\Delta m}{2}t}e^{\frac{\Delta\Gamma}{4}t}\right) \\ &= \frac{1}{4}e^{-\Gamma t}\left(e^{-\frac{\Delta\Gamma}{2}t} - e^{-i\Delta m t} + e^{i\Delta m t} - e^{\frac{\Delta\Gamma}{2}t}\right) \\ &= \frac{1}{2}e^{-\Gamma t}\left(-\sinh\frac{\Delta\Gamma}{2}t + i\sin\Delta m t\right) \end{aligned} \tag{3.33}$$

The time-dependent decay rate of a neutral meson produced as flavor eigenstate B to a final state f is given by

$$\frac{d\Gamma[B \to f](t)}{dt} = N_f|\langle f|B(t)\rangle|^2 \tag{3.34}$$

where N_f is a constant normalization factor. Using the definitions

$$A_f := \langle f|B\rangle \quad \text{and} \quad \bar{A}_f := \langle f|\bar{B}\rangle \tag{3.35}$$

and Eqs. (3.25), (3.32) and (3.33) the decay rate is given by

$$\begin{aligned}
\frac{d\Gamma[B\to f](t)}{dt} &= N_f \left| g_+(t)\langle f|B\rangle + \frac{q}{p} g_-(t)\langle f|\bar{B}\rangle \right|^2 \\
&= N_f \left(A_f^* g_+^*(t) + \bar{A}_f^* \frac{q^*}{p^*} g_-^*(t) \right) \left(A_f g_+(t) + \bar{A}_f \frac{q}{p} g_-(t) \right) \\
&= N_f \Bigg[|A_f|^2 |g_+(t)|^2 + A_f^* \bar{A}_f \frac{q}{p} g_+^*(t) g_-(t) \\
&\quad + \bar{A}_f^* A_f \frac{q^*}{p^*} g_-^*(t) g_+(t) + |\bar{A}_f|^2 \left|\frac{q}{p}\right|^2 |g_-(t)|^2 \Bigg] \\
&= \frac{1}{2} N_f e^{-\Gamma t} \Bigg[|A_f|^2 \left(\cosh \frac{\Delta\Gamma}{2} t + \cos \Delta m\, t \right) \\
&\quad + A_f^* \bar{A}_f \frac{q}{p} \left(-\sinh \frac{\Delta\Gamma}{2} t + i \sin \Delta m\, t \right) \\
&\quad + \bar{A}_f^* A_f \frac{q^*}{p^*} \left(-\sinh \frac{\Delta\Gamma}{2} t - i \sin \Delta m\, t \right) \\
&\quad + |\bar{A}_f|^2 \left|\frac{q}{p}\right|^2 \left(\cosh \frac{\Delta\Gamma}{2} t - \cos \Delta m\, t \right) \Bigg] \\
&= \frac{1}{2} N_f e^{-\Gamma t} \Bigg[\left(|A_f|^2 + \left| \bar{A}_f \frac{q}{p} \right|^2 \right) \cosh \frac{\Delta\Gamma}{2} t \\
&\quad + \left(|A_f|^2 - \left| \bar{A}_f \frac{q}{p} \right|^2 \right) \cos \Delta m\, t \\
&\quad - 2\mathrm{Re} \left(A_f^* \bar{A}_f \frac{q}{p} \right) \sinh \frac{\Delta\Gamma}{2} t - 2\mathrm{Im} \left(A_f^* \bar{A}_f \frac{q}{p} \right) \sin \Delta m\, t \Bigg].
\end{aligned} \tag{3.36}$$

With the definition

$$\lambda_f := \frac{\bar{A}_f}{A_f} \frac{q}{p} \tag{3.37}$$

Eq. (3.36) simplifies to

$$\begin{aligned}
\frac{d\Gamma[B\to f](t)}{dt} = N_f |A_f|^2 e^{-\Gamma t} \Bigg[&\frac{1+|\lambda_f|^2}{2} \cosh \frac{\Delta\Gamma}{2} t + \frac{1-|\lambda_f|^2}{2} \cos \Delta m\, t \\
&- \mathrm{Re}(\lambda_f) \sinh \frac{\Delta\Gamma}{2} t - \mathrm{Im}(\lambda_f) \sin \Delta m\, t \Bigg].
\end{aligned} \tag{3.38}$$

An analog calculation for the $\bar{B}$ state gives

$$\frac{d\Gamma[\bar{B} \to f](t)}{dt} = N_f |A_f|^2 \left|\frac{p}{q}\right|^2 e^{-\Gamma t} \left[\frac{1+|\lambda_f|^2}{2} \cosh \frac{\Delta\Gamma}{2} t - \frac{1-|\lambda_f|^2}{2} \cos \Delta m\, t \right. \left. -\mathrm{Re}(\lambda_f) \sinh \frac{\Delta\Gamma}{2} t + \mathrm{Im}(\lambda_f) \sin \Delta m\, t \right]. \tag{3.39}$$

Besides the term $|p/q|^2$, Eqs. (3.38) and (3.39) differ by the signs of the $\cos \Delta m\, t$ and $\sin \Delta m\, t$ terms. As we will see later, the former is exploited in measurements of Δm and the latter in CP violation measurements.

The decay rates to the CP conjugate final state, $|\bar{f}\rangle = CP|f\rangle$, are obtained by replacing A_f by $A_{\bar{f}} = \langle \bar{f}|B\rangle$ and $\bar{A}_f$ by $\bar{A}_{\bar{f}} = \langle \bar{f}|\bar{B}\rangle$ in Eqs. (3.38) and (3.39), implying λ_f is substituted by

$$\lambda_{\bar{f}} = \frac{\bar{A}_{\bar{f}}}{A_{\bar{f}}} \frac{q}{p}. \tag{3.40}$$

Because A_f and $A_{\bar{f}}$ are in general unrelated, but A_f and $\bar{A}_{\bar{f}}$ are related by the CP operation it is advisable to factor out the term $|\bar{A}_{\bar{f}}|^2$ instead of $|A_{\bar{f}}|^2$. Applying the above mentioned replacements to Eq. (3.38) yields:

$$\begin{aligned}
\frac{d\Gamma[B \to \bar{f}](t)}{dt} &= N_f |A_{\bar{f}}|^2 e^{-\Gamma t} \\
&\left[\frac{1+|\lambda_{\bar{f}}|^2}{2} \cosh \frac{\Delta\Gamma}{2} t + \frac{1-|\lambda_{\bar{f}}|^2}{2} \cos \Delta m\, t \right. \\
&\left. -\mathrm{Re}(\lambda_{\bar{f}}) \sinh \frac{\Delta\Gamma}{2} t - \mathrm{Im}(\lambda_{\bar{f}}) \sin \Delta m\, t \right] \\
&= N_f |A_{\bar{f}}|^2 |\lambda_{\bar{f}}|^2 e^{-\Gamma t} \\
&\left[\frac{|\lambda_{\bar{f}}|^{-2}+1}{2} \cosh \frac{\Delta\Gamma}{2} t + \frac{|\lambda_{\bar{f}}|^{-2}-1}{2} \cos \Delta m\, t \right. \\
&\left. -\mathrm{Re}(\lambda_{\bar{f}}/\lambda^*_{\bar{f}}\lambda_{\bar{f}}) \sinh \frac{\Delta\Gamma}{2} t - \mathrm{Im}(\lambda_{\bar{f}}/\lambda^*_{\bar{f}}\lambda_{\bar{f}}) \sin \Delta m\, t \right] \\
&= N_f |\bar{A}_{\bar{f}}|^2 \left|\frac{q}{p}\right|^2 e^{-\Gamma t} \\
&\left[\frac{1+|\lambda_{\bar{f}}|^{-2}}{2} \cosh \frac{\Delta\Gamma}{2} t - \frac{1-|\lambda_{\bar{f}}|^{-2}}{2} \cos \Delta m\, t \right. \\
&\left. -\mathrm{Re}(\lambda_{\bar{f}}^{-1}) \sinh \frac{\Delta\Gamma}{2} t + \mathrm{Im}(\lambda_{\bar{f}}^{-1}) \sin \Delta m\, t \right].
\end{aligned} \tag{3.41}$$

In an analog way one get from Eq. (3.39):

$$\begin{aligned}\frac{d\Gamma[\bar{B}\to\bar{f}](t)}{dt} =& N_f|\bar{A}_{\bar{f}}|^2 e^{-\Gamma t} \\ & \left[\frac{1+|\lambda_{\bar{f}}|^{-2}}{2}\cosh\frac{\Delta\Gamma}{2}t + \frac{1-|\lambda_{\bar{f}}|^{-2}}{2}\cos\Delta m\,t \right. \\ & \left. -\mathrm{Re}(\lambda_{\bar{f}}^{-1})\sinh\frac{\Delta\Gamma}{2}t - \mathrm{Im}(\lambda_{\bar{f}}^{-1})\sin\Delta m\,t\right]. \end{aligned} \tag{3.42}$$

We have set $N_{\bar{f}} = N_f$ because the normalization factor is determined by kinematics and identical for both final states.

The Cabibbo-favored weak decays of neutral mesons are often flavor specific, meaning the final state $f = f_{fs}$ is only accessible from B, but not from $\bar{B}$ mesons. An example are semileptonic decays where the charge of the lepton identifies the flavor of the initial quark. In flavor-specific decays $\bar{A}_f = A_{\bar{f}} = 0$ which implies $\lambda_f = \lambda_{\bar{f}}^{-1} = 0$ and the last two terms in Eqs. (3.38) to (3.42) that are caused by interference between B and $\bar{B}$ vanish. This leads to [8]

$$\frac{d\Gamma[B\to f_{fs}](t)}{dt} = \frac{1}{2}N_f|A_f|^2 e^{-\Gamma t}\left[\cosh\frac{\Delta\Gamma}{2}t + \cos\Delta m\,t\right] \tag{3.43}$$

$$\frac{d\Gamma[\bar{B}\to f_{fs}](t)}{dt} = \frac{1}{2}N_f|A_f|^2\left|\frac{p}{q}\right|^2 e^{-\Gamma t}\left[\cosh\frac{\Delta\Gamma}{2}t - \cos\Delta m\,t\right] \tag{3.44}$$

$$\frac{d\Gamma[B\to \bar{f}_{fs}](t)}{dt} = \frac{1}{2}N_f|\bar{A}_{\bar{f}}|^2\left|\frac{q}{p}\right|^2 e^{-\Gamma t}\left[\cosh\frac{\Delta\Gamma}{2}t - \cos\Delta m\,t\right] \tag{3.45}$$

$$\frac{d\Gamma[\bar{B}\to \bar{f}_{fs}](t)}{dt} = \frac{1}{2}N_f|\bar{A}_{\bar{f}}|^2 e^{-\Gamma t}\left[\cosh\frac{\Delta\Gamma}{2}t + \cos\Delta m\,t\right]. \tag{3.46}$$

Often decay rate asymmetries are studied. This has not only the advantage that some factors cancel out, but can also improve the precision of a measurement because systematic uncertainties cancel to some extent. A prime example is the asymmetry between flavor specific decays of not-oscillated and oscillated B mesons:

$$\mathscr{A}_{mix}(t) = \frac{d\Gamma[B\to f_{fs}](t)/dt - d\Gamma[B\to \bar{f}_{fs}](t)/dt}{d\Gamma[B\to f_{fs}](t)/dt + d\Gamma[B\to \bar{f}_{fs}](t)/dt}. \tag{3.47}$$

In case of $|A_f| = |\bar{A}_{\bar{f}}\; q/p|$ which is fulfilled if there is no CP violation in the decay and the mixing (see next Section) the mixing asymmetry is given by:

$$\mathscr{A}_{mix}(t) = \frac{\cos\Delta m\,t}{\cosh\frac{\Delta\Gamma}{2}t}. \tag{3.48}$$

This simplifies even more to $\mathscr{A}_{mix}(t) = \cos \Delta m\, t$ if the lifetime difference is negligible as it is the case for B^0 mesons. The mixing asymmetry is therefore well suited to measure the oscillation frequency Δm.

3.2 *CP* Violation

A violation of the *CP* symmetry is established by observing a difference between a process and the same process after applying the charge-conjugation and parity operation. The experimental signature is usually a non-vanishing rate asymmetry between both processes.

The origin of *CP* violation are complex terms in the Lagrangian that change the sign of their phase under *CP* transformation. In the standard model these phases are only present in *W* exchanges and are described by the CKM matrix. Therefore they are called weak phases. Another type of phases arises from hadronic interactions. There can be intermediate states that rescatter into the observed final state. Because these processes are dominated by strong interactions the involved phases are called strong phases. They are in general hard to calculate. Contrary to the weak phases they do not change their sign under *CP* conjugation. Both types of phases are convention dependent. Only phase differences have a physical meaning.

In order to violate the *CP* symmetry at least two terms with different weak phases are required. The interference between these terms determines the extent of the *CP* violation. In case of no interference *CP* is conserved.

Three classes of *CP* violating effects can be distinguished, *CP* violation in the decay, in mixing, and in the interference between the decay with and without mixing.

CP Violation in Decay

CP violation in the decay of a particle X is caused by different absolute values of the decay amplitudes for $X \to f$ and its *CP* conjugate $\bar{X} \to \bar{f}$:

$$\left| \frac{\bar{A}_{\bar{f}}}{A_f} \right| \neq 1. \tag{3.49}$$

Obviously, the magnitude of the *CP* violating effect depends on the considered decay mode. It is the only type of *CP* violation that can occur in weak decays of charged mesons and of baryons. However it can also be present in neutral meson decays. For non-oscillating particles it is experimentally measured by a time-independent decay rate asymmetry:

$$\mathscr{A}_{decay} = \frac{\Gamma[\bar{X} \to \bar{f}] - \Gamma[X \to f]}{\Gamma[\bar{X} \to \bar{f}] + \Gamma[X \to f]} = \frac{|\bar{A}_{\bar{f}}|^2 - |A_f|^2}{|\bar{A}_{\bar{f}}|^2 - |A_f|^2} = \frac{\left|\frac{\bar{A}_{\bar{f}}}{A_f}\right|^2 - 1}{\left|\frac{\bar{A}_{\bar{f}}}{A_f}\right|^2 + 1}. \quad (3.50)$$

CP violation in the decay is only possible if the decay amplitude has contributions from at least two diagrams. Then we can write it as

$$A_f = A_1 + A_2 = |A_1|e^{i(\delta_1+\phi_1)} + |A_2|e^{i(\delta_2+\phi_2)} \quad (3.51)$$

where $\delta_{1,2}$ and $\phi_{1,2}$ are the strong and weak phases, respectively, of the contributions $A_{1,2}$. *CP* conjugation yields

$$\bar{A}_{\bar{f}} = |A_1|e^{i(\delta_1-\phi_1)} + |A_2|e^{i(\delta_2-\phi_2)}. \quad (3.52)$$

For the determination of the decay rate asymmetry we have to calculate the absolute squared amplitudes:

$$\begin{aligned}
|A_f|^2 &= A_f^* A_f \\
&= \left(|A_1|e^{-i(\delta_1+\phi_1)} + |A_2|e^{-i(\delta_2+\phi_2)}\right)\left(|A_1|e^{i(\delta_1+\phi_1)} + |A_2|e^{i(\delta_2+\phi_2)}\right) \\
&= |A_1|^2 + |A_1||A_2|e^{i(\delta_2-\delta_1+\phi_2-\phi_1)} + |A_1||A_2|e^{-i(\delta_2-\delta_1+\phi_2-\phi_1)} + |A_2|^2 \\
&= |A_1|^2 + |A_2|^2 + 2|A_1||A_2|\cos(\delta_2-\delta_1+\phi_2-\phi_1) \\
&= |A_1|^2 + |A_2|^2 + 2|A_1||A_2|\,[\cos(\delta_2-\delta_1)\cos(\phi_2-\phi_1) \\
&\qquad - \sin(\delta_2-\delta_1)\sin(\phi_2-\phi_1)] \\
&= |A_1|^2 + |A_2|^2 + 2|A_1||A_2|\,[\cos\Delta\delta\cos\Delta\phi - \sin\Delta\delta\sin\Delta\phi] \quad (3.53)
\end{aligned}$$

where $\Delta\delta := \delta_2 - \delta_1$ and $\Delta\phi := \phi_2 - \phi_1$ are the strong and weak phase differences, respectively. With

$$|\bar{A}_{\bar{f}}|^2 = |A_1|^2 + |A_2|^2 + 2|A_1||A_2|\,[\cos\Delta\delta\cos\Delta\phi + \sin\Delta\delta\sin\Delta\phi] \quad (3.54)$$

the rate asymmetry becomes

$$\mathscr{A}_{decay} = \frac{2|A_1||A_2|\sin\Delta\delta\sin\Delta\phi}{|A_1|^2 + |A_2|^2 + 2|A_1||A_2|\cos\Delta\delta\cos\Delta\phi}. \quad (3.55)$$

This equation illustrates the requirements for *CP* violation in the decay. First of all there have to be at least two contributing diagrams. If A_2 is zero the rate asymmetry vanishes. Furthermore both terms must have different weak and strong phases. Fig. 3.1 illustrates this requirement.

From the experimental point of view a measurement of $\mathscr{A}_{decay}$ has the advantage that no time-dependent analysis is required and in particular for charged particles the

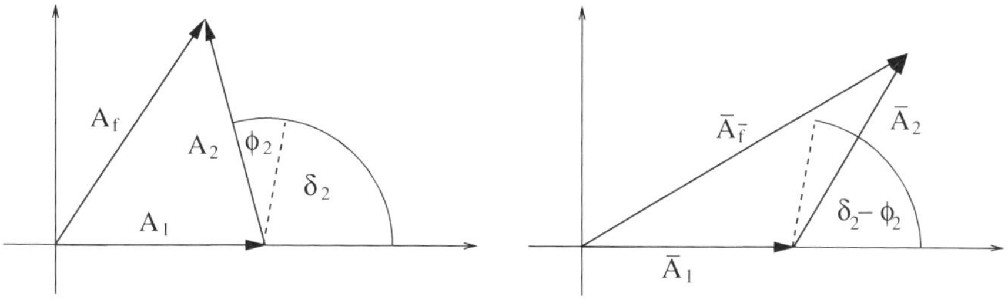

Fig. 3.1 Decay amplitude A_f composed of two contributions A_1 and A_2 (*left*) and its *CP* conjugate (*right*). Here the phase convention is chosen such that $\delta_1 = \phi_1 = 0$

two *CP* conjugate processes have well distinguishable signatures. But an extraction of the physically interesting weak phase difference that is related to the CKM matrix elements in the standard model is difficult. It requires the knowledge of the strong phase difference and of the ratio of absolute amplitudes $|A_2/A_1|$. This knowledge is often not available or only with limited precision.

CP Violation in Mixing

A *CP*-violating effect can be observed for neutral, oscillating mesons if their mass eigenstates, as defined in Eqs. (3.6) and (3.7), are not an equal mixture of their flavor eigenstates. This condition is expressed by the relation

$$\left|\frac{q}{p}\right| \neq 1 \quad \Leftrightarrow \quad |p| \neq |q|. \tag{3.56}$$

From Eqs. (3.6) and (3.7) follows that the two mass eigenstates are not orthogonal in this case:

$$\langle B_H|B_L\rangle = |p|^2 - |q|^2 \neq 0. \tag{3.57}$$

This also means that the mass eigenstates do not coincide with the *CP* eigenstates

$$|B_{even}\rangle = \frac{1}{\sqrt{2}}(|B\rangle + |\bar{B}\rangle) \quad \text{and} \quad |B_{odd}\rangle = \frac{1}{\sqrt{2}}(|B\rangle - |\bar{B}\rangle). \tag{3.58}$$

The deviation from the equal mixing is caused by a difference in the phase of M_{12} and Γ_{12}. The condition for *CP* violation in mixing can be related to the phase difference ϕ, defined in Eq. (3.56), using Eq. (3.16):

$$\left(\frac{q}{p}\right)^2 = \frac{\Delta m + i\Delta\Gamma/2}{2M_{12} - i\Gamma_{12}} \frac{2M_{12}^* - i\Gamma_{12}^*}{\Delta m + i\Delta\Gamma/2} = \frac{2M_{12}^* - i\Gamma_{12}^*}{2M_{12} - i\Gamma_{12}} \tag{3.59}$$

$$\Rightarrow \left|\frac{q}{p}\right|^4 = \left[\left(\frac{q}{p}\right)^2\right]^* \left(\frac{q}{p}\right)^2 = \frac{2M_{12} + i\Gamma_{12}}{2M_{12}^* + i\Gamma_{12}^*} \frac{2M_{12}^* - i\Gamma_{12}^*}{2M_{12} - i\Gamma_{12}}$$
$$= \frac{|M_{12}|^2 + |\Gamma_{12}|^2 + \mathrm{Im}(M_{12}\Gamma_{12}^*)}{|M_{12}|^2 + |\Gamma_{12}|^2 - \mathrm{Im}(M_{12}\Gamma_{12}^*)}$$
$$= \frac{|M_{12}|^2 + |\Gamma_{12}|^2 + |M_{12}||\Gamma_{12}|\sin\phi}{|M_{12}|^2 + |\Gamma_{12}|^2 - |M_{12}||\Gamma_{12}|\sin\phi} \tag{3.60}$$

We see that *CP* is conserved in mixing if $\sin\phi = 0$.

CP violation in mixing results in a difference between the rates of $B \to \bar{B}$ and $\bar{B} \to B$ transitions. It can be observed experimentally via an asymmetry in flavor specific decays that are only accessible via mixing, $B \to \bar{B} \to \bar{f}_{fs}$ and $\bar{B} \to B \to f_{fs}$:

$$\mathcal{A}CP_{mix}(t) = \frac{d\Gamma[\bar{B} \to f_{fs}](t)/dt - d\Gamma[B \to \bar{f}_{fs}](t)/dt}{d\Gamma[\bar{B} \to f_{fs}](t)/dt + d\Gamma[B \to \bar{f}_{fs}](t)/dt}. \tag{3.61}$$

Because often semileptonic decay modes are used to measure $\mathcal{A}CP_{mix}$ it is sometimes called semileptonic asymmetry, a_{SL}. In case of no *CP* violation in the decay ($|A_f| = |\bar{A}_{\bar{f}}|$) the asymmetry is, according to Eqs. (3.44) and (3.45), given by:

$$\mathcal{A}CP_{mix}(t) = \frac{\left|\frac{p}{q}\right|^2 - \left|\frac{q}{p}\right|^2}{\left|\frac{p}{q}\right|^2 + \left|\frac{q}{p}\right|^2} = \frac{1 - \left|\frac{q}{p}\right|^4}{1 + \left|\frac{q}{p}\right|^4}. \tag{3.62}$$

With Eq. (3.60) this becomes

$$\mathcal{A}CP_{mix}(t) = -\frac{|M_{12}||\Gamma_{12}|\sin\phi}{|M_{12}|^2 + |\Gamma_{12}|^2}. \tag{3.63}$$

This illustrates that although the decay rates are functions of time the asymmetry is time-independent. If $|\Gamma_{12}| \ll |M_{12}|$, as it it the case for B^0 and B_s^0 mesons, the decay rate asymmetry simplifies to

$$\mathcal{A}CP_{mix}(t) \approx -\frac{|\Gamma_{12}|}{|M_{12}|}\sin\phi. \tag{3.64}$$

As we have assumed $|\Gamma_{12}| \ll |M_{12}|$ the observed asymmetry should be small. With the approximations Eqs. (3.17) and (3.18) it transforms to

$$\mathcal{A}CP_{mix}(t) \approx -\frac{\Delta\Gamma}{\Delta m}\tan\phi. \tag{3.65}$$

So in order to extract the physically interesting phase ϕ we have to know $\Delta\Gamma$ and Δm.

CP Violation in Interference Between Mixing and Decay

The third type of *CP* violation is caused by the interference between decays with and without mixing. It requires a final state that is accessible from B and $\bar{B}$ mesons. Then the processes $B \to f$ and $B \to \bar{B} \to f$ can interfere and give access to the phase difference between both processes that leads to *CP* violation.

The condition for this type of *CP* violation, that is also called mixing induced *CP* violation, is given by

$$\mathrm{Im}(\lambda_f) \neq 0 \tag{3.66}$$

where λ_f is defined in Eq. (3.37) as $\lambda_f = (\bar{A}_f/A_f)(q/p)$. One can see that the condition depends on the decay amplitudes as well as on the mixing parameters p and q.

Experimentally, mixing induced *CP* violation can be observed by measuring the time-dependent asymmetry of B and $\bar{B}$ decays to a common final state f:

$$\mathcal{A}_{common}(t) = \frac{d\Gamma[\bar{B} \to f](t)/dt - d\Gamma[B \to f](t)/dt}{d\Gamma[\bar{B} \to f](t)/dt + d\Gamma[B \to f](t)/dt}. \tag{3.67}$$

With Eqs. (3.38) and (3.39) one obtains:

$$\begin{aligned}
\mathcal{A}_{common}(t) = \frac{1}{\Sigma}\Bigg[&\left|\frac{p}{q}\right|^2 \left(\frac{1+|\lambda_f|^2}{2}\cosh\frac{\Delta\Gamma}{2}t - \frac{1-|\lambda_f|^2}{2}\cos\Delta m\,t\right.\\
&\left.-\mathrm{Re}(\lambda_f)\sinh\frac{\Delta\Gamma}{2}t + \mathrm{Re}(\lambda_f)\sinh\frac{\Delta\Gamma}{2}t\right)\\
&-\left(\frac{1+|\lambda_f|^2}{2}\cosh\frac{\Delta\Gamma}{2}t + \frac{1-|\lambda_f|^2}{2}\cos\Delta m\,t\right.\\
&\left.-\mathrm{Re}(\lambda_f)\sinh\frac{\Delta\Gamma}{2}t - \mathrm{Re}(\lambda_f)\sinh\frac{\Delta\Gamma}{2}t\right)\Bigg]\\
= \frac{1}{\Sigma}\Bigg[&\left(\left|\frac{p}{q}\right|^2 - 1\right)\left(\frac{1+|\lambda_f|^2}{2}\cosh\frac{\Delta\Gamma}{2}t - \mathrm{Re}(\lambda_f)\sinh\frac{\Delta\Gamma}{2}t\right)\\
&-\left(\left|\frac{p}{q}\right|^2 + 1\right)\left(\frac{1-|\lambda_f|^2}{2}\cos\Delta m\,t - \mathrm{Im}(\lambda_f)\sin\Delta m\,t\right)\Bigg]
\end{aligned} \tag{3.68}$$

$$\begin{aligned}
\text{with}\quad \Sigma = &\left(\left|\frac{p}{q}\right|^2 + 1\right)\left(\frac{1+|\lambda_f|^2}{2}\cosh\frac{\Delta\Gamma}{2}t - \mathrm{Re}(\lambda_f)\sinh\frac{\Delta\Gamma}{2}t\right)\\
&-\left(\left|\frac{p}{q}\right|^2 - 1\right)\left(\frac{1-|\lambda_f|^2}{2}\cos\Delta m\,t - \mathrm{Im}(\lambda_f)\sin\Delta m\,t\right).
\end{aligned}$$

In case of no *CP* violation in mixing ($|q/p| = 1$) as expected to hold in good approximation for B^0 and B_s^0 mesons this simplifies to

$$\mathscr{A}_{common}(t) = \frac{(1-|\lambda_f|^2)\cos\Delta m\, t + 2\mathrm{Im}(\lambda_f)\sin\Delta m\, t}{(1+|\lambda_f|^2)\cosh\frac{\Delta\Gamma}{2}t + 2\mathrm{Re}(\lambda_f)\sinh\frac{\Delta\Gamma}{2}t}. \tag{3.69}$$

If in addition $\Delta\Gamma$ vanishes, as it is approximately the case for B^0 mesons, the expression reduces to

$$\begin{aligned}\mathscr{A}_{common}(t) &= \frac{1-|\lambda_f|^2}{1+|\lambda_f|^2}\cos\Delta m\, t + \frac{2\mathrm{Im}(\lambda_f)}{1+|\lambda_f|^2}\sin\Delta m\, t \\ &= C_f\cos\Delta m\, t + S_f\sin\Delta m\, t \end{aligned}\tag{3.70}$$

with

$$C_f := \frac{1-|\lambda_f|^2}{1+|\lambda_f|^2} \quad\text{and}\quad S_f := \frac{2\mathrm{Im}(\lambda_f)}{1+|\lambda_f|^2} \tag{3.71}$$

Of particular interest are final states that are *CP* eigenstates. One reason is that B and $\bar{B}$ mesons decay to them at approximately equal rate so that large interference effects are possible. Another advantage is that decay amplitudes can be related to each other via the *CP* operation. For a *CP* eigenstate one has

$$|\bar{f}_{CP}\rangle = CP|f_{CP}\rangle = \eta_f|f_{CP}\rangle \quad\text{with}\quad \eta_f = \pm 1 \tag{3.72}$$

where the sign is positive for *CP*-even and negative for *CP*-odd final states. With this $\bar{A}_{f_{CP}} = \eta_f\bar{A}_{\bar{f}_{CP}}$ and

$$\lambda_{f_{CP}} = \eta_f\frac{\bar{A}_{\bar{f}_{CP}}}{A_{f_{CP}}}\frac{q}{p}. \tag{3.73}$$

Then C_f becomes

$$C_{f_{CP}} = \frac{1-\left|\frac{\bar{A}_{\bar{f}_{CP}}}{A_{f_{CP}}}\right|^2}{1+\left|\frac{\bar{A}_{\bar{f}_{CP}}}{A_{f_{CP}}}\right|^2}. \tag{3.74}$$

Except for a sign, this is equal to the asymmetry between *CP* conjugated decay rates defined in Eq. (3.50). Thus the coefficient in front of the cos term in Eq. (3.70) measures the *CP* violation in the decay.

Now let us look at $S_{f_{CP}}$, the coefficient in front of the sin term, in case of a *CP*-even or -odd finale state. If in addition to no *CP* violation in mixing there is no *CP* violation in the decay ($|\bar{A}_{\bar{f}_{CP}}| = |A_{f_{CP}}|$), it is part of the only remaining term as $C_{f_{CP}}$ vanishes and Eq. (3.70) becomes

$$\mathscr{A}_{common}(t) = S_{f_{CP}} \sin \Delta m\, t \tag{3.75}$$

with

$$\begin{aligned} S_{f_{CP}} &= \mathrm{Im}(\lambda_{f_{CP}}) = \eta_f \mathrm{Im}\left(\frac{\bar{A}_{\bar{f}_{CP}}}{A_{f_{CP}}}\frac{q}{p}\right) \\ &= \eta_f \left|\frac{\bar{A}_{\bar{f}_{CP}}}{A_{f_{CP}}}\right| \left|\frac{q}{p}\right| \sin\left(\arg\left(\frac{\bar{A}_{\bar{f}_{CP}}}{A_{f_{CP}}}\frac{q}{p}\right)\right) \\ &= \eta_f \sin\left(\arg\left(\frac{q}{p}\right) + \arg(\bar{A}_{\bar{f}_{CP}}) - \arg(A_{f_{CP}})\right). \end{aligned} \tag{3.76}$$

As one can see from Eqs. (3.16) or (3.59) the phase of q/p is for $|\Gamma_{12}| \ll |M_{12}|$ given by the negative phase of M_{12}

$$\phi_M := \arg(M_{12}). \tag{3.77}$$

And since we have assumed no *CP* violation in the decay we can express $A_{f_{CP}}$ by a single strong and weak phase

$$A_{f_{CP}} = |A_{f_{CP}}|e^{i(\delta_f + \phi_f)} \quad \Rightarrow \quad \arg(A_{f_{CP}}) = \delta_f + \phi_f\ ,\ \arg(\bar{A}_{\bar{f}_{CP}}) = \delta_f - \phi_f. \tag{3.78}$$

Thus in the case of no *CP* violation in decay and mixing $S_{f_{CP}}$ is given by the mixing phase ϕ_M and the phase of the decay amplitude:

$$S_{f_{CP}} = -\eta_f \sin(\phi_M + 2\phi_f) \tag{3.79}$$

$$\Rightarrow \quad \mathscr{A}_{common}(t) = -\eta_f \sin(\phi_M + 2\phi_f) \sin \Delta m\, t. \tag{3.80}$$

As $S_{f_{CP}}$ depends only on weak phases and no hadronic terms enter, a measurement of this time-dependent asymmetry gives direct access to the physics parameters of interest, that are related to the phases of the CKM matrix elements.

Direct and Indirect CP Violation

Another way of classifying *CP* violating effects is the separation in direct and indirect *CP* violation. This classification was established after *CP* violation was observed for the first time in the neutral kaon system. Two different theories were able to describe this effect. Besides the Kobayashi-Maskawa theory, a theory of a superweak interaction [10] was discussed. This *CP* violating superweak force was postulated to change the flavor quantum number F (i.e. the strangeness S in the kaon system) by two units.

Therefore it was beneficial to characterize *CP* violating effects by the fact whether they could be explained by processes with $|\Delta F| = 2$ or they require processes with $|\Delta F| = 1$. The former case is called indirect *CP* violation and the latter one direct *CP* violation. Direct *CP* violation was first measured in $K \to \pi\pi$ decays [11–13] and disproved the superweak theory.

CP violation in the decay is caused by $|\Delta F| = 1$ processes and therefore belongs to the category of direct *CP* violation. As mixing is a $|\Delta F| = 2$ process, *CP* violation in mixing is indirect. A single measurement of *CP* violation in the interference between decays with and without mixing can be explained by a $|\Delta F| = 2$ process. Any observed asymmetry can be described with an appropriate choice of ϕ_M in Eq. (3.68). Evidence for direct *CP* violation is obtained if different asymmetries are measured for two different final *CP* eigenstates.

References

1. W. Fry, J. Schneps, M. Swami, Phys. Rev. **103**, 1904 (1956)
2. H. Albrecht et al., Phys. Lett. B**192**, 245 (1987)
3. C. Albajar et al., Phys. Lett. B**186**, 247 (1987)
4. A. Abulencia et al., Phys. Rev. Lett. **97**, 242003 (2006)
5. B. Aubert et al., Phys. Rev. Lett. **98**, 211802 (2007)
6. M. Staric et al., Phys. Rev. Lett. **98**, 211803 (2007)
7. T. Aaltonen et al., Phys. Rev. Lett. **100**, 121802 (2008)
8. K. Anikeev, et al., FERMILAB-PUB-01-197 (2001).
9. J. Beringer et al., Phys. Rev. **D86**, 010001 (2012)
10. L. Wolfenstein, Phys. Rev. Lett. **13**, 562 (1964)
11. H. Burkhardt et al., Phys. Lett. B**206**, 169 (1988)
12. V. Fanti et al., Phys. Lett. B**465**, 335 (1999)
13. A. Alavi-Harati et al., Phys. Rev. Lett. **83**, 22 (1999)

Chapter 4
Experimental Techniques

4.1 The Tevatron Collider

The Tevatron $p\bar{p}$ collider was operated by the Fermi National Accelerator Laboratory (Fermilab, FNAL) in Batavia near Chicago. It has two interaction points at which the experiments CDF and D0 detect the collision products. In the years 1992–1996 protons and antiprotons were accelerated to an energy of 0.9 TeV, leading to a center of mass energy of $\sqrt{s} = 1.8$ TeV. During this time, called Run I, both experiments collected about 0.1 fb^{-1} of data.

Afterwards the accelerator was upgraded in order to reach a center of mass energy of $\sqrt{s} = 1.96$ TeV. The Run II operation started in 2001 and ended on September 30th, 2011. Tevatron delivered an integrated luminosity of almost 12 fb^{-1}. Figure 4.1 shows the development of delivered luminosity versus time. It can be seen that the performance of the accelerator was steadily increased. This was mainly achieved by reaching higher instantaneous luminosities as shown in Fig. 4.2. The record peak luminosity is $4.14 \times 10^{32}\text{cm}^{-2}\,\text{s}^{-1}$.

To produce protons and antiprotons, and to accelerate them to an energy of nearly 1 TeV a sophisticated system of accelerator devices is needed. Figure 4.3 shows the involved components.

The accelerator chain starts with a Cockcroft-Walton pre-accelerator. It accelerates H^- ions from a magnetron surface plasma source in an electrostatic field to an energy of 750 keV. The hydrogen ions are fed to a linear accelerator which increases their energy to 400 MeV. Then the ions pass a carbon foil that strips off the electrons so that only protons remain. The protons are injected to the Booster, a ring accelerator with a circumference of 475 m. They leave the Booster, the last part of the so-called proton source, at an energy of 8 GeV.

The Main Injector is the next component in the pre-accelerator chain. It has a circumference of 3.3 km and has two tasks. First, it accelerates protons to an energy of 150 GeV for the injection into the Tevatron ring. Second, it delivers protons to fixed-target experiments and the antiproton source. In the latter case the protons are accelerated to only 120 GeV.

T. Kuhr, *Flavor Physics at the Tevatron*, Springer Tracts in Modern Physics 249,
DOI: 10.1007/978-3-642-10300-1_4,

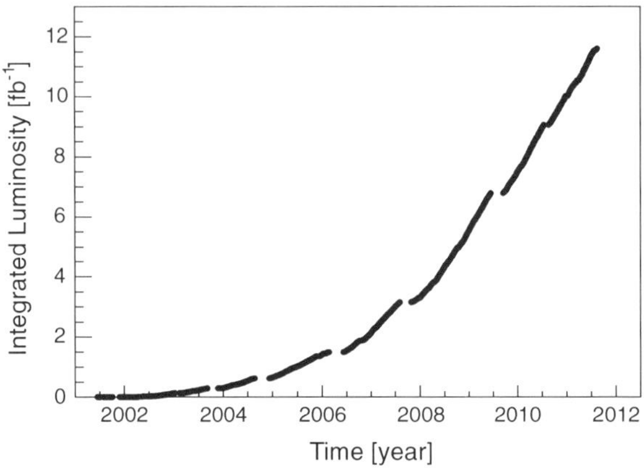

Fig. 4.1 Integrated luminosity delivered by the Tevatron in Run II [1]

Fig. 4.2 Peak luminosity delivered by the Tevatron in Run II [1]

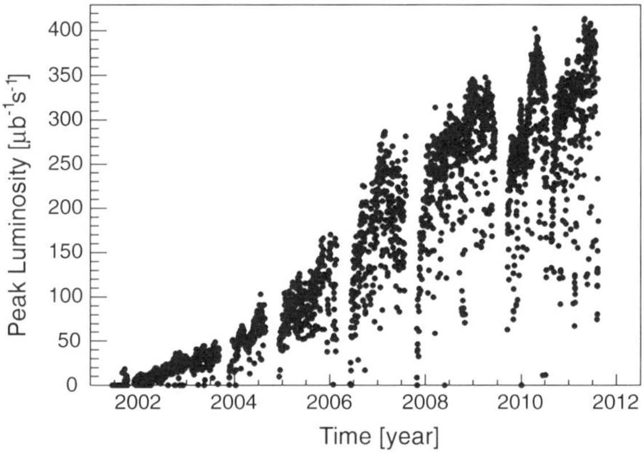

For the production of antiprotons the 120 GeV protons are shot on a nickel target. Among several other particles, antiprotons with the desired energy of about 8 GeV are created with an efficiency of the order of 10^{-5} per incoming proton. They are focused with a lithium lense and separated from the other particles with a mass spectrometer. In the Debuncher the spread in momentum is reduced by bunch rotation and stochastic cooling. Then the antiprotons are transferred to and collected in the Accumulator. Debuncher and Accumulator are triangle-shaped rings with a circumference of 505 and 474 m, respectively.

From the Accumulator, the last part of the so-called antiproton source, the particles are passed on to the Recycler, a storage ring in the Main Injector tunnel. It collects antiprotons at 8 GeV and provides stochastic [3] and electron cooling [4]. The pre-acceleration of the antiprotons to 150 GeV is performed by the Main Injector before they are injected into the Tevatron.

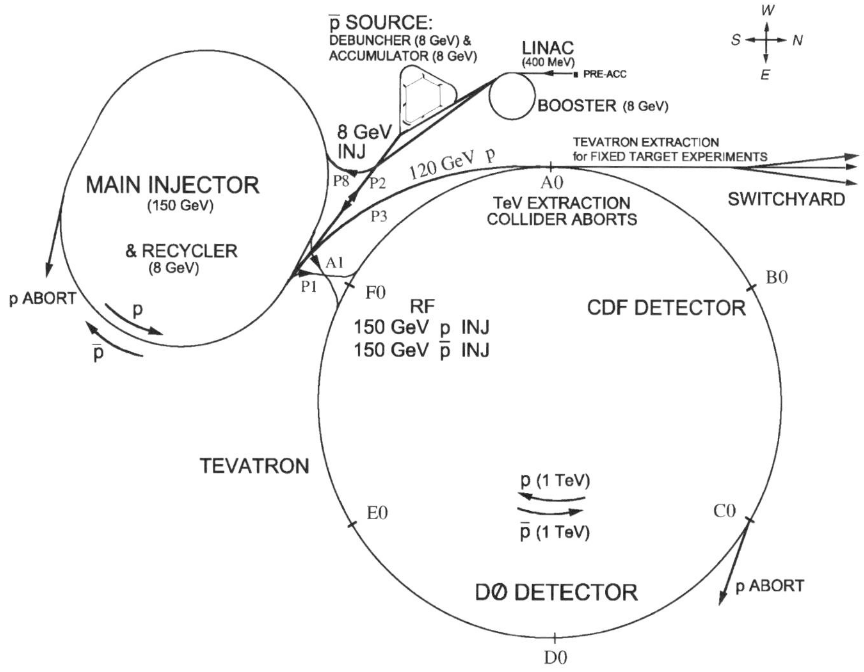

Fig. 4.3 Schematic view of the Tevatron collider and its pre-accelerators [2]

The Tevatron, a circular synchrotron with a radius of 1 km, then accelerates protons and antiprotons to the final energy of 980 GeV. The superconducting dipole magnets produce a field of up to 4.2 Tesla. The particles are concentrated in bunches that are about 120 m apart. This leads to a bunch crossing rate of 2.5 MHz, corresponding to one bunch crossing every 396 ns. However, since only 36 out of 53 possible bunches are filled the effective rate reduces to 1.7 MHz.

4.2 The CDF II Detector

The Collider Detector at Fermilab (CDF) [5–7] is a multi-purpose detector that measures the momenta and energies of particles produced in $p\bar{p}$ collisions. After Run I it was upgraded in order to cope with the increased interaction rate. Since then it is called CDF II detector. It is 12 m long, 12 m high, weighs 5 kT, and consists of several sub detectors that surround the interaction point.

The nominal interaction point defines the origin of the CDF coordinate system. The proton beam direction determines the direction of the z-axis. Polar angles with respect to the z-axis are denoted by the θ-coordinate. Alternatively the pseudorapidity η is often used to specify the direction relative to the beam axis. It is defined

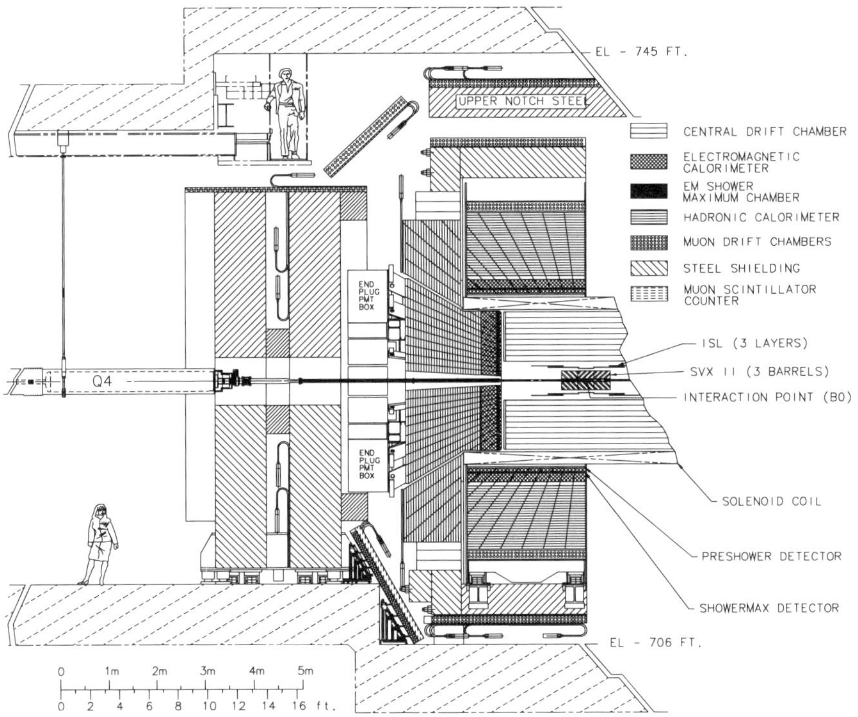

Fig. 4.4 Elevation view of one half of the forward-backward symmetric CDF II detector [7]

as $\eta = -\ln(\tan(\theta/2))$ and has an approximately flat distribution for particles produced within the detector acceptance. The x-axis is perpendicular to the beam direction, pointing away from the Tevatron center in horizontal direction. The y-axis points upwards so that an orthogonal right-handed coordinate system is defined. The momentum of particles projected on the xy-plane, the plane perpendicular to the beam axis, is called transverse momentum p_T. The angle in the transverse plane with respect to the x-axis is denoted by the azimuthal angle coordinate ϕ. The r-coordinate describes the radial distance from the z-axis.

The CDF II detector components have a cylindrical shape centered around the beam line. Figure 4.4 shows the individual components in an elevation view of one half of the forward-backward symmetric CDF II detector. The inner part consists of a tracking system that reconstructs the trajectory of charged particles. It is immersed in 1.4 T homogeneous magnetic field produced by a superconducting solenoid. Electromagnetic and hadronic calorimeters are located outside the tracking system and the magnet. They measure the energy of changed and neutral particles. Charged particles that are not absorbed by the calorimeters are detected by drift chambers in the outermost part of CDF II. Since almost only muons reach these sub detectors they are called muon chambers.

For flavor physics measurements the tracking system is an essential component. Charged particles are detected via ionization in a cylindrical drift chamber, the Central Outer Tracker (COT). It covers the range $|z| < 155$ cm and extends from a radius of 40 to 132 cm. Thus the acceptance for tracks reaching the outer part of the COT is the central region of $|\eta| < 1$. The chamber is radially divided into eight superlayers. Each superlayer is further divided into cells with a central plane of 12 sense wires. Grounded mylar field sheets define the cell boundaries. Because the electron drift direction differs from the electric field direction by the Lorentz angle of 35° the cells are rotated by this angle with respect to the radial direction. The superlayers are alternating of axial and stereo type. In an axial layer the sense wires are parallel to the beam axis and therefore measure the position in the transverse plane. Information about the z coordinate is obtained from stereo layers in which the cells are tilted by $\pm 2°$.

The reconstruction of tracks starts with the identification of track segments in each superlayer. Matching track segments are linked and a helix fit is performed. By using cosmic ray particles that are reconstructed in the COT as two tracks a transverse momentum resolution of $\sigma(p_T)/p_T^2 = 0.0015$ [c/GeV] was measured.

In addition to the trajectory of particles, the COT also measures their specific energy loss dE/dx. It is proportional to the logarithm of the charge deposited on a sense wire and determined by the pulse width of the read out signal. The specific energy loss depends on the momentum and mass of the particle and is described by a Bethe-Bloch formula [8]. For a particle with measured momentum it can thus be used to identify the type of particle (particle identification, PID). For example the average dE/dx values of pions and kaons with a momentum of 4 GeV/c are separated by about 1.5 standard deviations.

Besides the precise measurement of momenta flavor physics analyses often require the precise measurement of production and decay vertices. This is achieved by a silicon vertex detector inside the COT. It consists of three components, the Layer00 (L00), the Silicon Vertex Detector (SVX), and the Intermediate Silicon Layers (ISL). Layer00 are radiation hard micro strip sensors mounted directly on the beam pipe in two overlapping layers at radii of 1.35 and 1.62 cm. With a length of 94 cm it covers the range $|\eta| < 4$. The sensors have strips parallel to the beam axis on one side and thus measure the $r\phi$ coordinate in the transverse plane. The SVX consists of five layers of double-sided strip detectors. The innermost layer has a radius of 2.5 cm, the outermost of 10.6 cm yielding a pseudorapidity coverage of $|\eta| < 2$ for the 86 cm long detector. The strips on the two sides of layers 1, 2, and 4 are rotated by 90°. While one side measures the $r\phi$ coordinate the other one measures the z coordinate. Since multiple hits on a sensor lead to ambiguities for the linking of hits from both sides that can result in wrongly reconstructed space points, the strips on one side of layers 3 and 5 are rotated by only 1.2° with respect to the beam-axis-parallel strips on the other side (Small Angle Stereo, SAS). This reduced the ambiguities on the cost of a less precise measurement of the z coordinate. SAS sensors are used for the ISL, too. In the central region of $|\eta| < 1$ it consists of one layer at $r = 22$ cm. The forward regions $1 < |\eta| < 2$ are equipped with two layers at radii of 20 and 29 cm.

The reconstruction of tracks in the silicon detector starts with the extrapolation of tracks measured in the COT. In an iterative procedure going from the outermost layers to the inside hits are assigned and track parameters are refitted. Further track reconstruction algorithms search for tracks inside the silicon detector. By extrapolating these tracks in outward direction COT hits may be added to them. With the silicon vertex detector an impact parameter resolution of about 25 μm is achieved for high momentum tracks.

A further device for the identification of particles is the Time Of Flight detector (TOF) located between the COT and the magnet. It consists of 216 scintillator bars at a radius of 140 cm covering $|\eta| < 1$. The measurement of the arrival time of a particle together with the momentum and path length measured in the tracking system allows to determine the mass. With a resolution of about 0.1 ns a separation of pions and kaons of more than 2 standard deviations is achieved for momenta below 1.5 GeV/c.

The tracking system and the magnet are surrounded by sampling scintillator calorimeters arranged in a projective tower geometry. They are divided into an inner, electromagnetic part made of lead, and an outer, hadronic part made of steel. The central electromagnetic (CEM), central hadronic (CHA), end-wall hadronic (WHA), plug electromagnetic (PEM), and plug hadronic (PHA) calorimeters altogether cover a pseudorapidity range of $|\eta| < 3.6$.

While the energy measurement of particle jets in the calorimeters is essential for many analyses with high momentum objects, these components are less important for most flavor physics analyses. Here the main purpose is the identification of electrons. Electrons can be distinguished from hadrons by comparing the energy measured in the electromagnetic and hadronic calorimeter. Further information is provided by the Central Electromagnetic Shower detector (CES) and the Central Pre-Radiator (CPR). The CES consists of proportional chambers embedded in the CEM that measure the shower profile at the expected maximum for electrons. The CPR chambers provide a measurement in front of the CEM.

With a depth of 5.5 interaction lengths the central calorimeters provide a good shielding for the muon chambers. The muon chambers detect charged particles in proportional chambers and scintillators. The Central Muon Detector (CMU) has a coverage of $|\eta| < 0.6$. Particles must have a transverse momentum p_T of at least 1.4 GeV/c to reach the CMU. To reduce the rate of hadrons misidentified as muon a second component is placed around the CMU and the magnetic return yoke. The Central Muon Upgrade (CMP) covers the same pseudorapidity range of $|\eta| < 0.6$. To reach this box-shaped detector a p_T of at least 2 GeV/c is required. The conical sections of the Central Muon Extension (CMX) increase the acceptance up to $|\eta| = 1$. The Barrel Muon Chambers (BMU) are mounted around the forward toroids that are not used in Run II and cover $1 < |\eta| < 1.5$.

The identification of muons starts with the recognition and reconstruction of track segments in the muon chambers. Hits in at least three out of four layers are required. A track reconstructed in the tracking system is called a muon candidate if its extrapolation to the muon chambers can be matched to a muon track segment.

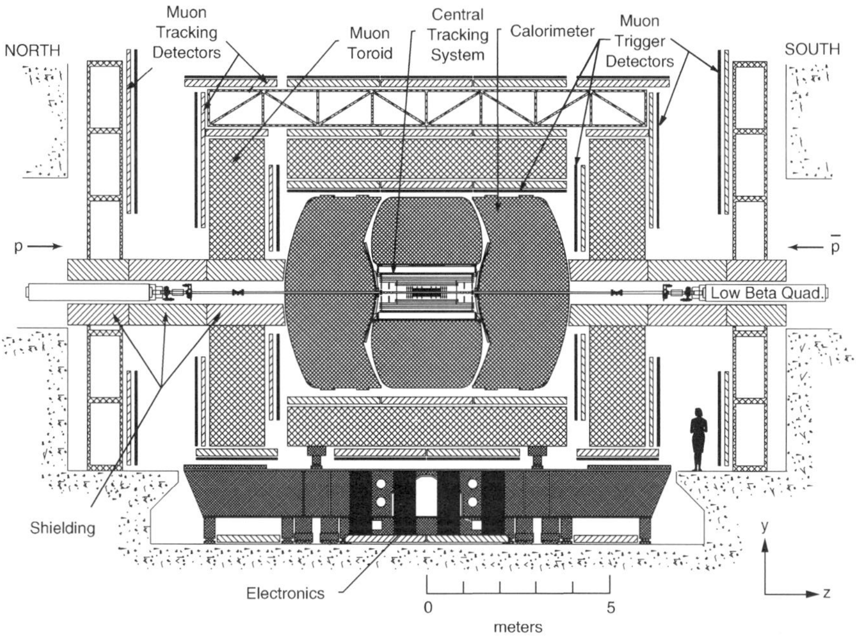

Fig. 4.5 Elevation view of the D0 detector [9]

Cherenkov Luminosity Counters (CLC) are mounted close to the beam pipe on each side of the detector. By measuring the rate of inelastic $p\bar{p}$ collisions the luminosity can be determined with an accuracy of about 6 %.

4.3 The D0 Detector

The D0 detector [9], named after the Tevatron section where it is located, has a length of 15 m, a height of 9 m, and weighs 5 kT. It is a multi-purpose detector like the CDF II detector. Although the general layout of both detectors is similar different detection technologies and configurations with different strengths and weaknesses allow to obtain complementary results and to cover a large field of physics analyses. While the D0 detector was optimized for measurements of high momentum objects in Run I the upgrade for Run II improved its flavor physics capabilities.

The coordinate system is defined in the same way as the CDF one. The individual components are illustrated in Fig. 4.5. The innermost part is the tracking system that is located inside a superconducting solenoid. The magnet produces a 2 T homogeneous field for the momentum measurement of charged particles. Tracking system and solenoid are surrounded by electromagnetic and hadronic calorimeters with preshower detectors. Muon detectors are the outermost components. They are placed inside and outside of toroidal magnets.

The reconstruction of charged particle tracks starts with the detection of hits in the Silicon Microstrip Tracker (SMT). Four layers of silicon sensors from $r = 2.7$ to 10.5 cm are arranged in six barrels around the beam pipe. All sensors have strips parallel to the beam axis on one side providing a measurement of the $r\phi$ coordinate. In addition the layers 2 and 4 have strips with a stereo angle of 2° on the second side allowing to determine the z position of hits. The central four barrels of layers 1 and 3 have strips perpendicular to the beam axis on the second side yielding a more precise z coordinate measurement on the cost of potential $r\phi$ and z hit matching ambiguities. Between each of the 12 cm long barrels and at the end of the barrels so-called F-disks are mounted. The twelve F-disks are perpendicular to the beam axis and consist of twelve wedges of double-sided silicon sensors with a stereo angle of 30°. Their radial dimension goes from 2.6 to 10.0 cm yielding a coverage of $|\eta| < 3$. Further on in forward direction two more disks were placed. These so-called H-disks consist of single sided sensors at $r = 9.5 - 20$ cm. They were removed in 2006 at the end of the data taking period called Run IIa in order to have space for the readout cables of a new layer of silicon sensors in Run IIb. The added Layer-0 silicon detector consists of single sided sensors at a radius of 1.6 cm arranged in a hexagonal shape around the beam pipe. It is divided in eight barrels. The central four have a length of 7 cm, the outer four a length of 12 cm. Layer-0 improves the track impact parameter resolution by about 30 %.

The second part of the tracking system is the Central Fiber Tracker (CFT). It consists of eight cylindrical dual layers of scintillating polystyrene fibers with a diameter of 836 μm. The inner one of each of the two layers is aligned parallel to the beam direction and called axial layer. The second layer, called stereo layer, is tilted by $+3°$ or $-3°$. The innermost dual layer at a radius of 20 cm has a length of 1.66 m. The outermost fibers at a radius of 52 cm cover with a length of 2.52 m a pseudorapidity range of $|\eta| < 1.7$. The photons that are produced by charged particles traversing the scintillating fibers are transmitted via clear fiber waveguides to visible light photon counters (VLPC). They are locate in a cryostat below the calorimeter and detect single photons with an efficiency of about 80 % leading to a hit efficiency of $\sim$99.9 %.

The energy of electrons, photons and hadrons is measured in sampling calorimeters surrounding the tracking system and the solenoid. The Central Calorimeter (CC) has an acceptance of $|\eta| < 1$. The two End Calorimeters (EC) extend the acceptance to $|\eta| \approx 4$. Each of the calorimeters is divided into an electromagnetic, a fine hadronic, and a coarse hadronic part. The absorber material is mainly uranium for the first two types and copper (stainless steel) for the coarse hadronic CC (EC). As active material liquid argon is used. Scintillator strips are mounted on the inner surface of the calorimeter cryostats. These preshower detectors help to identify electrons.

Like the calorimeters the Muon System is divided into a central part (CF) and a forward part on the north and south side (EF). The central muon system is shielded by the 7.2 interaction lengths of the central calorimeter and covers the pseudorapidity range $|\eta| < 1.5$. The forward muon system extends the coverage to $|\eta| < 2.0$. Each of the muon systems consists of a toroidal magnet and three layers of drift tubes, one inside and two outside of the toroid. The magnetic field inside the toroids has an average strength of about 1.8 T and allows for a measurement of the momentum independent of the measurement in the central tracking system. A special feature of the D0 detector is that the field direction of the solenoid and the toroids is inverted regularly after several days. This helps to control systematic uncertainties due to detector asymmetries.

Protons and anti-protons scattered elastically under a small angle of the order of 1 mrad can be identified by the Forward Proton Detector (FPD). It consists of scintillating fibers that can be moved close to the beam line several meters away from the interaction point. Like at CDF the instantaneous luminosity is measured by detecting inelastic $p\bar{p}$ reactions. The Luminosity Monitor (LM) uses plastic scintillation counters mounted close to the beam pipe at $z = \pm 1.4$ m.

4.4 Triggering of Heavy Flavor Events

The effective bunch crossing rate of 1.7 MHz is much too high to read out, record and analyze each of these events. With a raw data event size of about 150 kB it would correspond to a data rate of more than 200 GB per second. To reduce this rate to a manageable size a tight selection of events has to be applied. Moreover the selection procedure, called triggering, has to be fast to cope with an input rate of 2.5 MHz.

The decision which events are kept and which are discarded is motivated by the physics processes that are of interest for analysis. They usually have a much lower rate than the dominant process of inelastic $p\bar{p}$ scattering. As both collaborations have a broad physics program different selection requirements have to be considered. This can make it a challenging task to find a good compromise that utilizes the delivered luminosity in an optimal way. The design and the tuning of the trigger setup plays a crucial role for the potential to extract physics results from the recorded data. Actually, one of the main reasons why some analyses are performed by one experiment and not by the other one are the different trigger capabilities.

Both Tevatron experiments employ a three level trigger system for the online event selection. The first level, L1, is implemented in hardware. Because it needs more time than the 0.4 ns between bunch crossings for a decision the events are stored in a buffer. The information available at this level comes from the calorimeters, the muon system, and the central tracker (COT or CFT, respectively). Requirements on the transverse momentum, charge and azimuthal angle of tracks can be applied. A coarse muon identification is achieved by matching tracks in the central tracker to signals in the muon system. The L1 output rate is 1.5 kHz at D0 and up to 40 kHz at CDF. The accepted events are written to a second buffer and passed to the next trigger level.

At the second level, L2, that is implemented in hardware, too, information from the silicon vertex detectors (SVX and SMT, respectively) becomes available. This allows to refine the track parameters determined by L1 and to measure the impact parameter with respect to the primary vertex. The L2 output rate is of the order of 500 Hz for CDF and 800 Hz for D0.

On the third trigger level, L3, software algorithms run on PC farms. Here the full detector information is available. This allows for a confirmation of the L1 and L2 decision using more precisely determined parameters and for the application of more sophisticated selection criteria. If an event was selected by L3 it is written to a disk buffer and then stored on tape. The final output rate is about 100 Hz for both experiments.

In order to select heavy flavor events from inelastic scattering events which have a three orders of magnitude higher cross section than $b\bar{b}$ production several characteristic features of heavy flavored hadrons are exploited. A well identifiable signature are charmonium and bottomonium states decaying to a pair of muons. In addition muon pairs can arise from rare decays, like $B_s^0 \rightarrow \mu^+\mu^-$, or semileptonic decays of two heavy hadrons in the same event. Both experiments trigger efficiently on pairs of muons. In order to reduce background these triggers usually pose requirements on the invariant mass of the pairs. In the beginning of Run II when the instantaneous luminosity was low CDF triggered also on decays of charmonia to e^+e^-.

Another class of events are those with only one muon from a semileptonic decay of a heavy hadron. D0 triggers on single muons with a transverse momentum of at least 3 GeV/c. As the rate of this trigger becomes too high at high luminosities a tighter version of it with a requirement of $p_T > 5$ GeV/c is added. While the D0 detector and trigger are optimized for muon identification CDF has to require an additional track displaced from the primary vertex in order to trigger on semileptonic decays. This second track from the heavy hadron decay vertex is identified by an impact parameter with respect to the primary vertex of 120 μm $< d_0 <$ 1 mm reconstructed at L2 by the Silicon Vertex Trigger (SVT). The so-called SVT track has to have a transverse momentum of at least 2 GeV/c. The lepton can be an electron or muon and must have $p_T > 4$ GeV/c.

The most challenging signature to trigger on at the Tevatron are the hadronic decays of heavy hadrons. The handle to distinguish them from background is the displaced decay vertex. Therefore CDF triggers on L1 on a pair of tracks with $p_T >$ 2 GeV/c each. On L2 both tracks are required to be well matched to hits in the silicon detector and their impact parameter has to be in the range of 120 μm $< d_0 <$ 1 mm. A further reduction of background from tracks, that come from the primary vertex, is achieved by a cut on the distance between the primary vertex and the intersection point of both tracks in the transverse plane of $L_{xy} > 200$ μm. In addition the scalar sum of the two transverse momenta, $p_{T,1} + p_{T,2}$, has to be above a given threshold. Three different versions of this trigger with thresholds at 4, 5.5, and 6.5 GeV/c are used. At the third level of the so-called Two-Track-Trigger (TTT) the requirements are verified with improved precision. Since D0 has no track trigger on L1 it can only acquire events with hadronic heavy flavor decays by triggering on muons from the semileptonic decay of a second heavy hadron in the same event from $c\bar{c}$ or $b\bar{b}$ pair production.

The different versions of a trigger, that mainly differ in their minimal momentum requirement, are necessary because the interaction rates and therefore the trigger rates

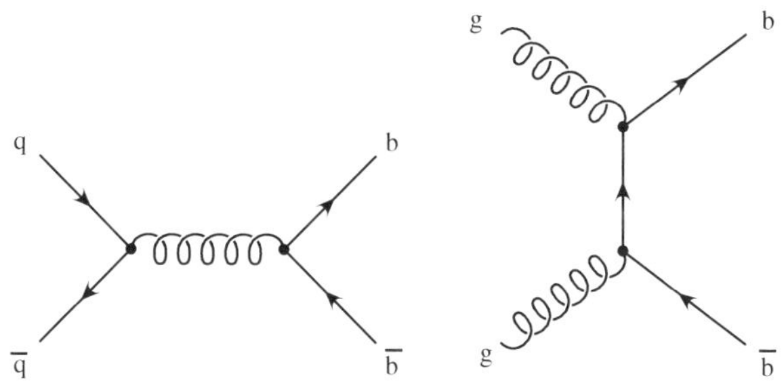

Fig. 4.6 Production of a $b\bar{b}$ pair in the hard scattering process (flavor creation)

change with time. This includes not only the decrease of instantaneous luminosity during a Tevatron store, but also the improvements in the initial instantaneous luminosity achieved by the accelerator operators over the past years. One way to exploit, but not exceed, the available output bandwidth is the use of luminosity dependent trigger settings. Another option to fit the trigger rate to the available bandwidth is to throw away some part of the events that were triggered. In this method, known as prescaling, a number $n \geq 1$ is assigned to a trigger and then only each nth event fulfilling the trigger requirement is passed on. In this way the trigger rate can be easily adjusted to the available bandwidth, but of course it reduces the effective integrated luminosity by the prescale factor n. In the analysis of the recorded data the changes in the trigger system with time, via luminosity dependent settings or prescales, can have severe consequences and may require careful consideration.

4.5 Simulation of Heavy Flavor Events

In order to extract physical parameters from the measured data, often a detailed understanding of the detector response to physical processes is needed. This is usually obtained by simulating such processes and the interaction of the produced particles with the detector. Because (pseudo) random numbers are extensively used in this procedure it is referred to as Monte Carlo (MC) simulation.

A frequently used event generator for heavy flavor events at hadron colliders is Pythia [10]. It generates the four-momenta of quarks and gluons according to leading order QCD calculations. Effects from higher order processes are emulated by initial and final state parton showers. The momenta of the initial partons are taken from parametrized parton distribution functions [11].

A large amount of heavy flavor events can be generated easily if Pythia is set to generate a heavy quark pair in the hard scattering process as shown in Fig. 4.6. However, this production mechanism, called flavor creation, accounts only for part of the total heavy flavor cross section. Other contributions come from heavy sea quarks in the proton or antiproton that take part in the hard scattering process. This

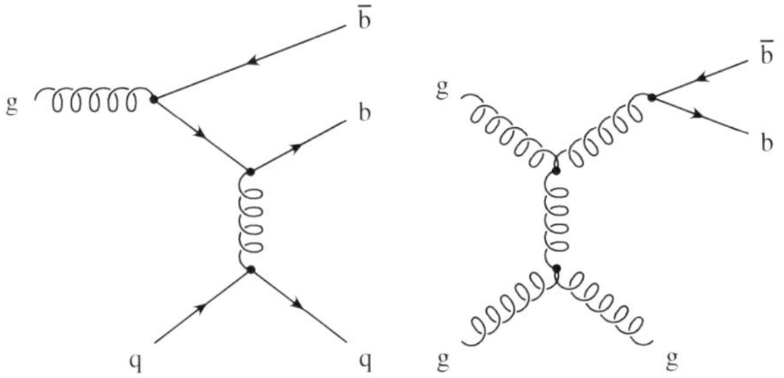

Fig. 4.7 Production of a b quarks via flavor excitation (*left*) and gluon splitting (*right*)

production mechanism is called flavor excitation and depicted in Fig. 4.7 on the left side. The Feynman diagram next to it shows the third possible production mechanism, gluon splitting, where the heavy quark pair is created from a hard gluon. To get a MC sample that accurately describes data, all three processes have to be taken into account. This is achieved by generating events with all quark flavors in the hard process and then selecting the events with heavy quarks in the final state. Because this is a very time consuming procedure such samples are often reused for several analyses.

Pythia does not only simulate the hard scattering process, but also the fragmentation of partons into hadrons and the underlying event which results from the remaining partons, that do not participate in the hard interaction. Thus the Pythia simulation is intended to describe all features of real data events.

In many analyses this level of comprehension is not needed because only the properties of the heavy hadron are of interest. Then it is sufficient to generate just this one hadron and no underlying event. In this case the kinematic distribution of the heavy hadron is taken from measured spectra.

Although Pythia provides code for the decay of heavy hadrons, the EvtGen package [12] is used instead in case of B mesons. The package is provided by the BaBar collaboration and contains a more sophisticated model of B meson decays. For simulating the radiation of photons in the final state the Photos package [13] is used.

The interaction of the produced particles with the detector is simulated by GEANT3 [12]. After a detector-specific simulation of the digitization and trigger response the MC data is available in the same format as real data.

References

1. Fermilab Accelerator Division (2012). http://www-bd.fnal.gov/
2. M.A. Pleier, Int. J. Mod. Phys. **A24**, 2899 (2009)
3. S. Van Der Meer, Rev. Mod. Phys. **57**, 689 (1985)
4. V. Shiltsev, V. Danilov, D. Finley, A. Seryi, Phys. Rev.ST Accel. Beams **2**, 071001 (1999)
5. D.E. Acosta et al., Phys. Rev. **D71**, 032001 (2005)

6. D.E. Acosta et al., Phys. Rev. **D71**, 052003 (2005)
7. A. Abulencia et al., J. Phys. **G34**, 2457 (2007)
8. J. Beringer et al., Phys. Rev. **D86**, 010001 (2012)
9. V.M. Abazov et al., Nucl. Instrum. Meth. **A565**, 463 (2006)
10. T. Sjostrand et al., Comput. Phys. Commun. **135**, 238 (2001)
11. H.L. Lai et al., Eur. Phys. J. **C12**, 375 (2000)
12. D.J. Lange, Nucl. Instrum. Meth. **A462**, 152 (2001)
13. P. Golonka, Z. Was, Eur. Phys. J. **C45**, 97 (2006)

Chapter 5
Lifetime and Decay of B-Hadrons

The lifetimes and branching ratios of weakly decaying b hadrons are important parameters for two reasons. First, they serve as benchmarks for theoretical predictions. Second, they enter many other measurements, either directly as input parameters of fits to data, or indirectly as parameters of MC simulations.

In the spectator model, where the quarks in the hadrons are treated as independent, non-interacting particles, the lifetime of ground state hadrons consisting of a b and one or two light quarks is determined by the weak decay of the b quark. Thus all these b hadrons would have equal lifetime, neglecting the tiny contribution from the weak decay of the s quark.

But the experimentally observed differences between the lifetime of heavy hadrons shows that the interaction of the heavy quark with the spectator quarks gives significant contributions to the decay width. A case where the spectator quark matters is shown in Fig. 5.1. The right diagram is color suppressed because colorless final state hadron can only be formed if the color of the quarks from the W decay matches the color of the c quark and the u spectator quark. But since it has the same final state as the color allowed tree level process shown in the left part of Fig. 5.1, both can interfere and thus lead to a significant contribution. Note, that such an interference, called Pauli interference, does not occur for B^0 mesons because the corresponding diagrams have different final states as shown in Fig. 5.2. Since the interference is destructive this leads to a longer lifetime of the B^+ than the B^0 meson. This effect is even more pronounced for charm mesons where the ratio of D^+ to D^0 lifetimes is about 2.5.

Further decay processes involving the spectator quark are weak annihilation and weak exchange, illustrated in Fig. 5.3. While the former is only possible for charged B mesons, the latter contributes only to neutral B mesons and baryons.

Theoretical calculation of heavy hadron lifetimes are often based on heavy quark expansion (HQE) [1–5]. In this approach the total decay width is determined in an expansion in inverse heavy quark mass m_Q. At first order the naive spectator model prediction of equal lifetimes is obtained. Higher order corrections lead to the prediction of the following hierarchy of lifetimes: $\tau(B^+) > \tau(B^0) \approx \tau(B_s^0) > \tau(\Lambda_b^0) \gg \tau(B_c^+)$. Often ratios of lifetimes are considered because several uncertainties in the

T. Kuhr, *Flavor Physics at the Tevatron*, Springer Tracts in Modern Physics 249,
DOI: 10.1007/978-3-642-10300-1_5, © Springer-Verlag Berlin Heidelberg 2013

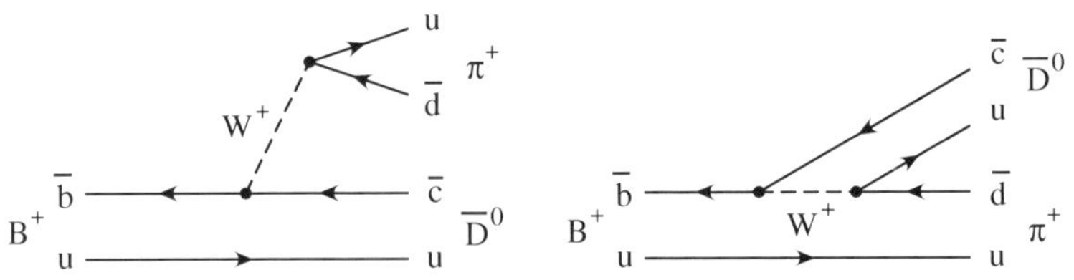

Fig. 5.1 Feynman diagrams of $B^+ \to \bar{D}^0\pi^+$ decay that contribute to Pauli interference

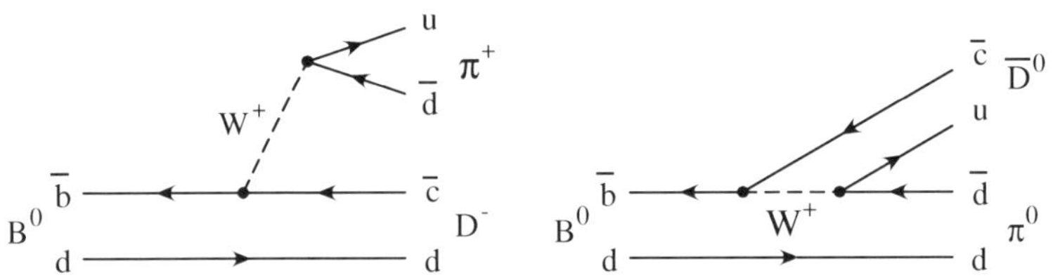

Fig. 5.2 Feynman diagrams corresponding to Fig. 5.1 for the B^0 decay

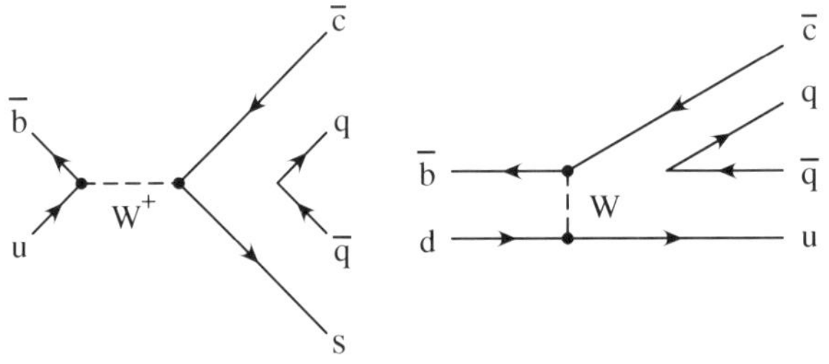

Fig. 5.3 Feynman diagrams for weak annihilation (*left*) and weak exchange (*right*)

theoretical calculation cancel in this case. Also from the experimental point of view ratios can usually be determined more precisely because of the cancellation of systematic uncertainties.

In this chapter we will first discuss measurements of the B^0 and B^+ lifetimes which can be regarded the standard candles for b hadron lifetimes and are important parameters for the simulation of B meson decays. The average lifetime and the decay width difference of B_s^0 mesons are in particular relevant for measurements of CP violation in the B_s^0 system. Finally, lifetime measurements of B_c^+ mesons and Λ_b^0 baryons allow to study B hadron decays in a different environment of two heavy quarks and a system with a spectator di-quark, respectively. Partial decay widths measurements provide additional information about the individual modes contributing to the total decay rate. Several B_s^0 decay modes are observed for the first time at the Tevatron and open windows to measure further physical parameters.

5.1 B^0 and B^+

As a consequence of the different trigger categories at the Tevatron experiments, different methods are employed to measure the B^0 and B^+ lifetimes.

One of them uses decays to J/ψ mesons, $B^+ \to J/\psi K^+$, $B^0 \to J/\psi K^*(892)^0$, and $B^0 \to J/\psi K^0_S$ [6]. An advantage of this method is that the events are recorded by a $J/\psi \to \mu^+\mu^-$ trigger whose efficiency is independent of the B meson decay time. The observed decay time distribution is thus given by an exponential convolved with a resolution function.

In this analysis by the CDF collaboration the B meson decay vertex, $\mathbf{x}_{SV}$, is determined by the vertex of the two muons from the $J/\psi \to \mu^+\mu^-$ decay. Since the kaon is not used in the vertex reconstruction, the position resolution is independent of the mode. To ensure a precise vertex determination, at least three hits in the silicon detector are required for each muon track.

The primary vertex position, $\mathbf{x}_{PV}$, is determined from the average interaction region of the proton and anti-proton beams. This region, called beamline, is obtained from a fit of primary tracks collected over several events and has a width of about 30 μm in the xy plane. Because the beamline is tilted with respect to the z-axis, the beamspot position at the z-coordinate of the J/ψ vertex is taken as primary vertex.

The B meson decay time in the meson's rest frame, t, can then be calculated from the distance between the primary and J/ψ decay vertex, and the reconstructed B meson momentum. But in this approach the decay time resolution is dominated by the spacial resolution in z direction. Thus, to improve the resolution, the measurement is restricted to the transverse plane. Furthermore, the flight distance vector from the primary to the secondary vertex is projected on the B meson momentum direction:

$$L_{xy} = (\mathbf{x}_{SV} - \mathbf{x}_{PV}) \cdot \frac{\mathbf{p}_T}{p_T}, \tag{5.1}$$

where L_{xy} is the transverse decay length, $\mathbf{p}_T$ the B meson momentum in the transverse plane, and $p_T = |\mathbf{p}_T|$. Since the uncertainty of the momentum direction is much smaller than that of the flight direction, this results in an additional improvement in decay time resolution. The proper decay time in the B meson rest frame is calculated as

$$t = \frac{L_{xy}}{v_T \gamma} = \frac{M L_{xy}}{p_T}, \tag{5.2}$$

where v_T is the velocity in the transverse plane, γ is the relativistic factor, and M is the B meson mass.

The projection on the B meson momentum can lead to negative L_{xy} values if the reconstructed flight and momentum vectors point in opposite directions. Such events are useful to study resolution effects. In fully reconstructed events, as used in this analysis, the resolution is given by the vertex reconstruction uncertainty. The contribution from the momentum uncertainty is negligible.

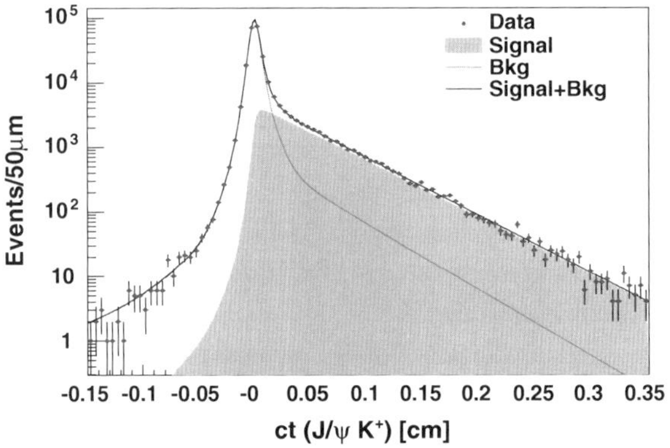

Fig. 5.4 Decay time distribution and fit projection for the $B^+ \to J/\psi K^+$ sample [6]

Selection requirements on particle momenta, vertex fit qualities, the pointing angle between the B flight and momentum direction, and reconstructed $K^{*0} \to K^+\pi^-$ and $K_S^0 \to \pi^+\pi^-$ masses are optimized to obtain a maximal signal significance $S/\sqrt{S+B}$, where S is the signal yield estimated from simulation and B the background yield estimated from mass sidebands.

The average lifetimes are determined in an unbinned maximum likelihood fit (see Fig. 5.4). The fitted observables are the reconstructed invariant mass and decay time and their measured uncertainties. The event-by-event uncertainties enter in the resolution functions of the decay time and the signal mass, respectively. Since signal and background have different decay time uncertainty distributions they are taken into account in the fit by empirical models.

Because the decay time resolution is not perfectly Gaussian, but has outliers, the resolution function is modeled by the sum of three Gaussians with mean zero and different widths. Each width is given by the measured decay time uncertainty times an individual scale factor. These scale factors and the relative fractions of the Gaussians are determined from a fit to mass sideband data that contains mainly prompt events.

The dominant systematic uncertainty is the alignment of the silicon detector. But, because it is correlated between the modes, it cancels in the ratio of lifetimes. Further systematic uncertainties arise from the models used in the probability density function (PDF).

In a dataset corresponding to an integrated luminosity of 4.3 fb^{-1} CDF measures $\tau(B^+) = [1.639 \pm 0.009\,(\text{stat}) \pm 0.009\,(\text{syst})]$ ps, $\tau(B^0) = [1.507 \pm 0.010\,(\text{stat}) \pm 0.008\,(\text{syst})]$ ps, and $\tau(B^+)/\tau(B^0) = 1.088 \pm 0.009\,(\text{stat}) \pm 0.004\,(\text{syst})$ [6]. The result is compared with other measurements in Fig. 5.7.

Another approach to measure lifetimes is based on semileptonic decays which has the advantage of a higher branching ratio compared to methods based on decays into charmonium states. This method is used by D0 to measure the ratio of B^+ to B^0 lifetimes in a data sample corresponding to an integrated luminosity of 440 pb^{-1} [7].

Two inclusive decay modes are reconstructed, $B \to D^{*-}\mu^+ X$ and $B \to \bar{D}^0\mu^+ X$ where X stands for any decay products that are not explicitly reconstructed, including the neutrino. The first mode is dominated by B^0 decays ($89 \pm 3\,\%$), the second by B^+ decays ($83 \pm 3\,\%$).

The D^0 mesons are reconstructed from $K^-\pi^+$ pairs and the D^{*+} via its decay to $D^0\pi^+$. Only if a D^0 candidate cannot be combined with a pion to a D^{*+} candidate, it is used in the $B \to \bar{D}^0\mu^+ X$ sample. A sample of D^{*+} mesons combined with μ^+ tracks of same charge is used to estimate the contribution of combinatorial background.

A drawback of semileptonic decays is the incomplete reconstruction of the B meson. Because the momentum of the B meson is only partially reconstructed, Eq. (5.2) cannot be used to determine the proper decay time. Instead, a visible proper decay time is defined:

$$t_{vis} = \frac{M(\mathbf{x}_{SV} - \mathbf{x}_{PV}) \cdot \mathbf{p}(\,\mathrm{D}^{(*)}\mu)_T}{p(D^{(*)}\mu)_T^2}, \tag{5.3}$$

where $\mathbf{p}(\mathrm{D}^{(*)}\mu)_T$ is the reconstructed momentum of the $D^{(*)}\mu$ combination in the transverse plane.

The difference between the visible and true proper decay time is described by a factor $K = p(D^*\mu)_T/p(B)_T$ so that $t = Kt_{vis}$. Its distribution, $D(K)$, is determined from simulation taking into account all known contributions to the reconstructed final state. The distribution of visible proper decay times is then given by a combination of exponentials with decay times τ/K:

$$P(t_{vis}) = \int dK\; D(K)\theta(t_{vis})\frac{K}{\tau}\exp\left(-\frac{Kt_{vis}}{\tau}\right) \tag{5.4}$$

where θ is the step function.

Using such a relation for each sample and taking into account resolution and efficiency effects, the ratio of B^+ to B^0 lifetimes is determined in a fit to the ratio of yields in bins of t_{vis} as shown in Fig. 5.5. Systematic uncertainties are due to the yield determination procedure using fits to mass distributions, the efficiency, resolution model and K factor distribution. The obtained lifetime ratio is $\tau(B^+)/\tau(B^0) = 1.080 \pm 0.016\,(\text{stat}) \pm 0.014\,(\text{syst})$ [7].

Finally, hadronic B decays, that are selected by CDF's Two-Track-Trigger, can be used for lifetime measurements. Since these decays are usually fully reconstructed, they have a good decay time resolution comparable with the resolution of decays to charmonium states. The challenge for these modes is the decay time dependent trigger acceptance that is caused by the requirements on impact parameters and decay lengths.

One way to deal with this effect is to correct for it using an acceptance function derived from simulation. Since the systematic uncertainties in this approach are hard to control CDF has developed a new method that allows to measure the lifetime

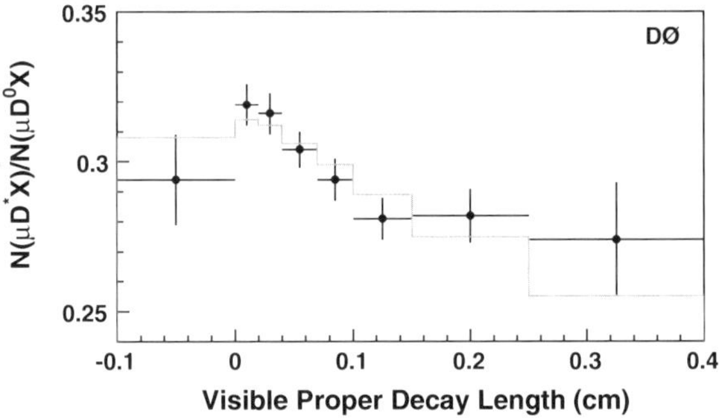

Fig. 5.5 Ratio of signal yields in the $B \rightarrow D^{*-}\mu^+X$ and $B \rightarrow \bar{D}^0\mu^+X$ samples (points with error bars) as a function of ct_{vis} compared with the fit projection (histogram) [7]

of a B meson without relying on MC simulation [8] and applied it to the decay $B^+ \rightarrow \bar{D}^0\pi^+$ with $\bar{D}^0 \rightarrow K^+\pi^-$.

The basic idea is to use an event dependent acceptance function. For an event with a given decay kinematics, the impact parameters of the daughter particles can be calculated as a function of the B meson decay time t. So for each daughter particle track one can determine in which time range it satisfies the trigger requirements. Since the Two-Track-Trigger requires two tracks within a certain impact parameter window, one can then obtain the trigger acceptance for the B meson. Basically one takes each observed B meson decay, and virtually shifts its decay time to determine at which times the observed decay kinematics fulfills the trigger requirements.

While the impact parameter at trigger level is available in the offline analysis, a complication arises from the fact that the track finding efficiency on trigger level is lower than in the offline reconstruction. Their ratio is approximately constant within the trigger range. Since the reconstructed decay mode has three tracks in the final state, the trigger efficiency can have four different values. It is maximal if all three tracks have impact parameters within the trigger range. If only two out of three tracks fulfill the trigger requirements the efficiency is lower. It drops to zero if only one or no track is within the trigger acceptance. Figure 5.6 illustrates the determination of the event dependent trigger efficiency.

With the event dependent efficiency and a description of the detector resolution one can then define the decay time PDF for signal events. The decay time distribution of background events is parametrized by an empirical model. Another complication in this analysis is caused by the difference in efficiency functions between signal and background. As explained in Ref. [9], if an event-by-event variable is used in a likelihood function and the distribution of this variable is different for different components, then the result of a likelihood fit is biased if the PDFs of the variable are not taken into account.

In this analysis the event dependent observable is the efficiency function. It was shown with pseudoexperiments that neglecting the difference in the efficiency functions between signal and background leads to a bias in the fitted lifetime. The

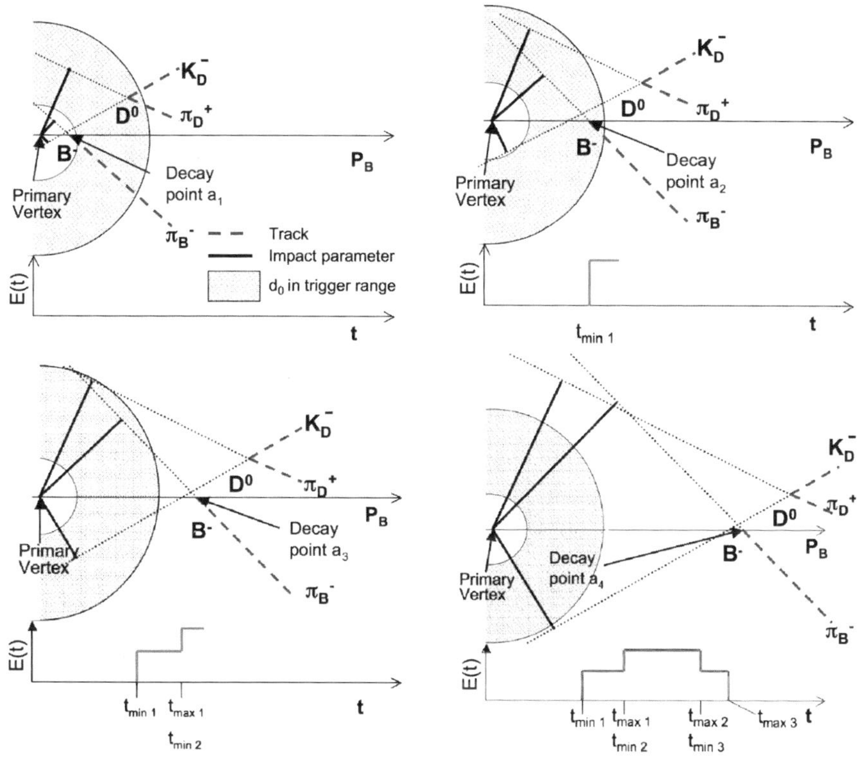

Fig. 5.6 Illustration of decay time dependent trigger efficiency, $E(t)$, for a $B^- \to D^0\pi^-$ decay with given kinematics [8]. In the *upper left plot* only one track has an impact parameter within the trigger range, indicated by the two circles for the minimal and maximal impact parameter requirement, so that the event is not selected and the efficiency is zero. In the *upper right* and *lower left* plot two and three tracks fulfill the trigger requirement, respectively. Because the track finding efficiency on trigger level is not 100 %, the probability that the trigger will select the event is higher in the latter case. When the impact parameters of the tracks exceed the trigger range the efficiency drops to zero again as shown in the *lower right plot*

challenge in this case is that not distributions of real number values for signal and background are needed to cure the effect, but distributions of functions.

To deal with this challenge a projection from the space of functions to a real number is introduced. The projection uses a Fisher discriminant that optimizes the separation between signal and background in the multidimensional space of parameters describing the efficiency functions. By including the PDFs of the Fisher discriminant output in the likelihood function an unbiased result is obtained.

The measured B^+ lifetime is 1.663 ± 0.023 (stat) ± 0.015 (syst) ps [8] where the dominant systematic uncertainties come from the track finding efficiency on trigger level and a possible correlation between mass and decay time in the background.

As one can see in Fig. 5.7, the result is well consistent with the world average, indicating that the analysis technique is reliable. The comparison plot also shows good

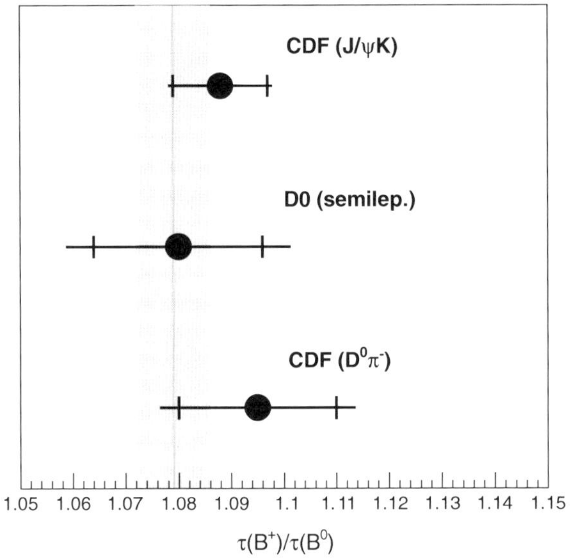

Fig. 5.7 Comparison of measured B^+ to B^0 lifetime ratios [6–8], where the last one is calculated from the measured $\tau(B^+)$ value and the world average B^0 lifetime [10]. The band shows the world average ratio [10] including these measurements

agreement between the measurements in $J/\psi K$ and semileptonic final states. Overall the individual results obtained with different analysis techniques and affected by different systematic uncertainties yield a consistent picture. The CDF measurement in the $J/\psi K$ channel is the most precise single measurement today.

5.2 B_s^0 Decays

While a variety of B^0 and B^0 decay modes are observed and measured at the B factories, several B_s^0 decays were first observed at the Tevatron. Some of these modes are particularly relevant for measurements of *CP* violation in the B_s^0 system.

Since the production rate of B_s^0 mesons at the Tevatron is not well known, measurements of branching ratios are performed relative to a normalization mode. Often a B^0 decay mode, that is related to the B_s^0 decay via the exchange of s and d quarks, called U spin symmetry, is taken for the relative branching ratio measurement. On the one hand this has the advantage, that some theoretical uncertainties cancel in the calculation of the branching fraction ratio. Another advantage is the fact that the kinematics of both decays is very similar and several experimental systematic uncertainties are reduced, too.

One of the experimentally best accessible B_s^0 decay modes is $B_s^0 \to D_s^-\pi^+$. It is a Cabibbo-favored tree level decay and in contrast to the corresponding $B^0 \to D^-\pi^+$

decay it has no contribution from an exchange diagram (cf. left diagram in Fig. 5.3). Its branching ratio was measured by CDF with data sample of 115 pb^{-1} in 2006 [11]. The result was updated one year later using 355 pb^{-1} [12]. In this analysis also the relative branching ratio $\mathcal{B}(B_s^0 \to D_s^- \pi^+\pi^+\pi^-)/\mathcal{B}(B^0 \to D^0\pi^+\pi^+\pi^-)$ was measured for the first time. Both modes were essential for the first observation of B_s^0 oscillation.

The relative branching ratios are measured using the following equation

$$\frac{\mathcal{B}(B_s^0 \to D_s^- \pi^+[\pi^+\pi^-])}{\mathcal{B}(B^0 \to D^0\pi^+[\pi^+\pi^-])} = \frac{N(B_s^0)}{N(B^0)} \frac{\varepsilon(B^0)}{\varepsilon(B_s^0)} \frac{f_d}{f_s} \frac{\mathcal{B}(D^-)}{\mathcal{B}(D_s^-)}, \tag{5.5}$$

where $N(B_s^0)$ and $N(B^0)$ are the numbers of reconstructed B_s^0 and B^0 signal events, and $\varepsilon(B_s^0)$ and $\varepsilon(B^0)$ are the corresponding reconstruction and selection efficiencies, respectively. The factor f_d/f_s is the ratio of production rates of B^0 to B_s^0 mesons. The terms $\mathcal{B}(D^-)$ and $\mathcal{B}(D_s^-)$ represent the branching ratios of the D^- and D_s^- decays used in the analysis. In this case these are $D^- \to K^+\pi^-\pi^-$ and $D_s^- \to \phi\pi^-$ with $\phi \to K^+K^-$, $D_s^- \to K^{*0}K^-$ with $K^{*0} \to K^+\pi^-$, or $D_s^- \to \pi^+\pi^+\pi^-$.

The events are selected online by the Two-Track-Trigger. The signal yields are determined in a fit to the invariant $D_s^-\pi^+[\pi^+\pi^-]$ and $D^-\pi^+[\pi^+\pi^-]$ mass distributions. The relative efficiency is taken from simulation. The relative production rate and the branching ratios of the daughter particles are external inputs.

Using a value of $f_s/f_d = 0.258 \pm 0.038$ [13] the relative branching ratios $\mathcal{B}(B_s^0 \to D_s^-\pi^+)/\mathcal{B}(B^0 \to D^0\pi^+) = 1.13 \pm 0.08\,(\text{stat}) \pm 0.05\,(\text{syst}) \pm 0.23\,(\text{norm})$ and $\mathcal{B}(B_s^0 \to D_s^-\pi^+\pi^+\pi^-)/\mathcal{B}(B^0 \to D^0\pi^+\pi^+\pi^-) = 1.05 \pm 0.10\,(\text{stat}) \pm 0.07\,(\text{syst}) \pm 0.21\,(\text{norm})$ are obtained [12], where the last uncertainty is due to the normalization on the corresponding B^0 decay. A nice byproduct of the analysis is the finding that the dominant contribution in the three pion system of the $B_s^0 \to D_s^-\pi^+\pi^+\pi^-$ decay is the a_1 resonance as one can see by the peaking structure at around 1.2 GeV in Fig. 5.8.

While large samples of $B_s^0 \to D_s^-\pi^+[\pi^+\pi^-]$ decays were collected at the Tevatron, rarer B_s^0 decays became accessible. One interesting mode is the Cabibbo-suppressed $B_s^0 \to D_s^\mp K^\pm$ decay. As can be seen from the Feynman diagrams in Fig. 5.9, the decays to both final state charge combinations are possible and have a comparable amplitude. Via B_s^0 mixing both processes can interfere, leading to a sizable *CP* violating effect. The phase between the two diagrams is $\gamma = \arg(-V_{ud}V_{ub}^*/V_{cd}V_{cb}^*)$ (cf. Eq. (2.12)), the angle of the unitarity triangle that is least constrained so far.

An experimental challenge in the measurement of the angle γ via this decay is the need to identify the B_s^0 flavor at production and to resolve the fast B_s^0 oscillations in a time dependent measurement. Another issue, that has to be resolved first, is the separation of the $B_s^0 \to D_s^\mp K^\pm$ signal from the Cabibbo-favored and kinematically similar $B_s^0 \to D_s^-\pi^+$ decay. The latter was achieved first by CDF with a data sample corresponding to an integrated luminosity of 1.2 fb^{-1} [14].

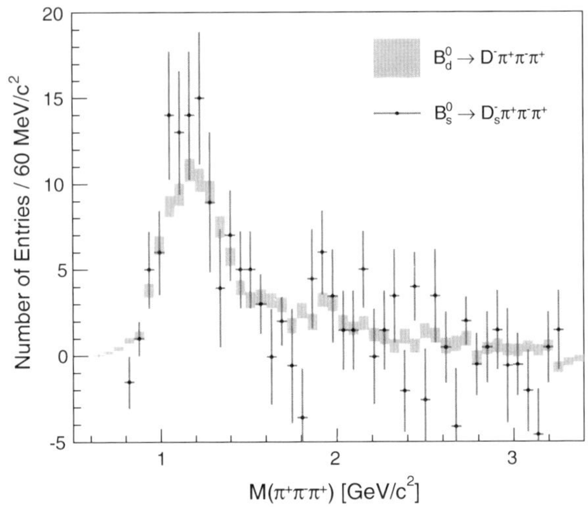

Fig. 5.8 Comparison of three pion invariant mass distributions for $B_s^0 \to D_s^- \pi^+ \pi^+ \pi^-$ and $B^0 \to D^0 \pi^+ \pi^+ \pi^-$ decays [12]

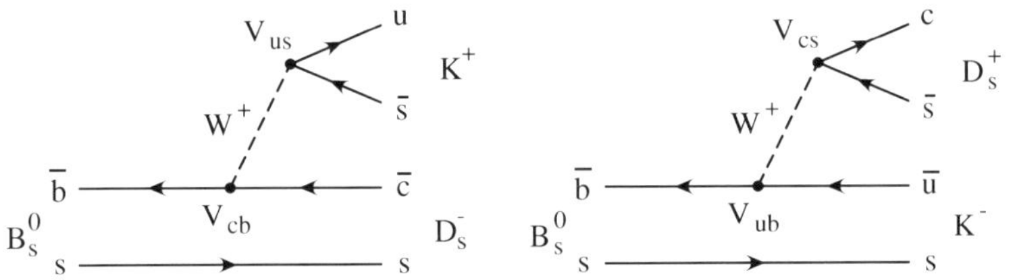

Fig. 5.9 Leading order Feynman diagrams of $B_s^0 \to D_s^- K^+$ (*left*) and $B_s^0 \to D_s^+ K^-$ decays (*right*)

The D_s^- meson is reconstructed in the decay to $\phi\pi^-$. B_s^0 candidates are formed from combinations of D_s^- candidates and charged particle tracks. To statistically separate the signal from the background, a fit to the B_s^0 candidate invariant mass, $m(D_s^-\pi^+)$ and the variable Z, which is the logarithm of the ratio between the measured and the expected dE/dx in the drift chamber for the B_s^0 daughter track. In the calculation of both variables the pion hypothesis is assumed. Thus the background contribution from $B_s^0 \to D_s^-\pi^+$ decays is expected to peak at the nominal B_s^0 mass in $m(D_s^-\pi^+)$ and at zero in Z, while the signal component is shifted to lower values in both variables.

The shape of the radiative tail of $B_s^0 \to D_s^-\pi^+$ decays, which overlaps with the signal mass distribution, is taken from a PHOTOS simulation [15] and its normalization is floating in the fit. Other backgrounds are partially reconstructed b hadron decays, $B \to DX$, and combinatorial background. The Z PDFs for kaons and pions are obtained from a high statistics $D^{*+} \to D^0(\to K^-\pi^+)\pi^+$ sample.

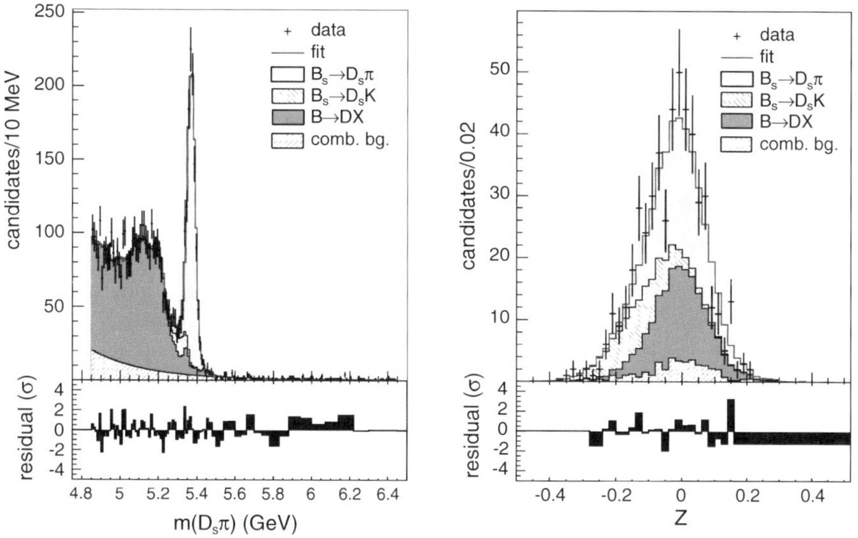

Fig. 5.10 Fit projections on $m(D_s^- \pi^+)$ (*left*) and Z (*right*) compared with the data distributions [14]

The fit projections are shown in Fig. 5.10. The analysis method was verified on $B^0 \to D^{(*)-}K^+/\rho^+/\pi^+$ control samples.

The observed $B_s^0 \to D_s^{\mp}K^{\pm}$ signal has a statistical significance of 8.1 standard deviations. The branching ratio is measured relative to the normalization mode $B_s^0 \to D_s^-\pi^+$ to be $\mathcal{B}(B_s^0 \to D_s^{\mp}K^{\pm})/\mathcal{B}(B_s^0 \to D_s^-\pi^+) = 0.097 \pm 0.018$ (stat) $\pm$ 0.009 (syst) [14], where the dominant systematic uncertainties come from the uncertainties of the Z and mass PDFs.

Semileptonic B_s^0 decays have been studied at the Tevatron as well. As such decays are often used, either inclusively or in exclusive modes, to measure the size of CKM matrix elements, it is important to understand which exclusive modes contribute what fraction to the inclusive decay rate.

The decay $B_s^0 \to D_{s1}^-(2536)\mu^+X$, where $D_{s1}^-(2536)$ is an orbitally excited state of the D_s^- meson with angular momentum $L = 1$ and $J^P = 1^+$, was first observed by D0 in a data sample corresponding to an integrated luminosity of 1.3 fb^{-1} [16]. The D_{s1}^- meson is reconstructed in the decay to $D^{*-}K_S^0$ with $D^{*-} \to \bar{D}^0(\to K^+\pi^-)\pi^-$ and $K_S^0 \to \pi^+\pi^-$. In events with a D_{s1}^- and a muon candidate, the yield of B_s^0 mesons is extracted from a fit to the $m(D^{*-}K_S^0)$ mass spectrum as shown in Fig. 5.11.

The production rate times branching ratio product is measured relative to $b \to D^{*-}\mu^+X$ events and yields $f_s \cdot \mathcal{B}(B_s^0 \to D_{s1}^-(2536)\mu^+X) \cdot \mathcal{B}(D_{s1}^-(2536) \to D^{*-}K_S^0) = [2.66 \pm 0.52\ (\text{stat}) \pm 0.45\ (\text{syst})] \times 10^{-4}$ [16]. The systematic error is dominated by the uncertainties on the efficiencies of the K_S^0 and $D^{*-}\mu^+$ selection.

Finally also B_s^0 decays to charmonium states are studied at the Tevatron. The "golden" mode for measurements of CP violation in the B_s^0 system is $B_s^0 \to J/\psi\,\phi$

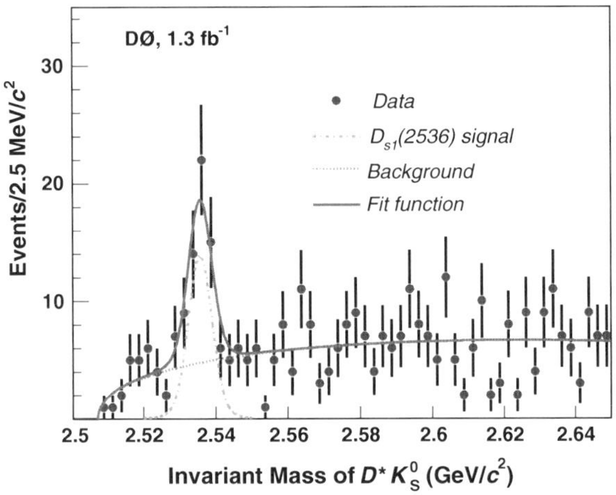

Fig. 5.11 Fit projection and data distribution of $D^{*-}K^0_S$ invariant mass [16]

and will be discussed in Chap. 7. Using the same analysis technique, the decay $B^0_s \to \psi(2S)\phi$ could be used for an independent CP violation measurement.

This B^0_s decay mode was first observed by CDF using 360 pb^{-1} [17] and later confirmed by D0 using 1.3 fb^{-1} of data [18]. Both experiments reconstruct $\psi(2S)$ mesons in dimuon decays, in addition CDF uses $\psi(2S) \to J/\psi\pi^+\pi^-$ events. The ϕ is reconstructed from pairs of oppositely charged kaons. The mode $B^0_s \to J/\psi\phi$ is used for normalization, where $J/\psi \to \mu^+\mu^-$.

The signal yields are determined from fits to invariant mass spectra (see Fig. 5.12). The measured relative branching ratios $\mathcal{B}(B^0_s \to \psi(2S)\phi)/\mathcal{B}(B^0_s \to J/\psi\phi)$ are 0.52 ± 0.13 (stat) $\pm$ 0.04 (syst) $\pm$ 0.06 ($\mathcal{B}$) for CDF [17] and 0.53 ± 0.10 (stat) $\pm$ 0.07 (syst)$\pm$0.06 ($\mathcal{B}$) for D0 [18], where the last uncertainty comes from the J/ψ and $\psi(2S)$ branching ratios. In both cases the source of the largest systematic uncertainty is the unknown polarization of $B^0_s \to \psi(2S)\phi$ decays.

Cabibbo-suppressed decays to charmonium states are next in the shopping list of searches for new B^0_s decay modes. The decays $B^0_s \to J/\psi K^{*0}$ with $K^{*0} \to K^+\pi^-$ and $B^0_s \to J/\psi K^0_S$ with $K^0_S \to \pi^+\pi^-$ were first observed by CDF in a data sample of 5.9 fb^{-1} [19]. The mode $B^0_s \to J/\psi K^{*0}$ is a decay of a pseudoscalar to a pair of vector mesons, like the golden mode decay $B^0_s \to J/\psi\phi$, and could help to control the hadronic uncertainties in it [20]. The $J/\psi K^0_S$ final state is CP-odd and thus would allow to directly measure the lifetime of the CP-odd B^0_s eigenstate. While this would already provide important information on the decay width difference in the B^0_s system and thus contribute to the search for new physics, a sufficiently large data sample would allow to measure the CP-violating phase without the need to perform an angular analysis.

The selection for both modes is optimized using the expected signal yields from simulation and background expectation from mass sideband extrapolations. While

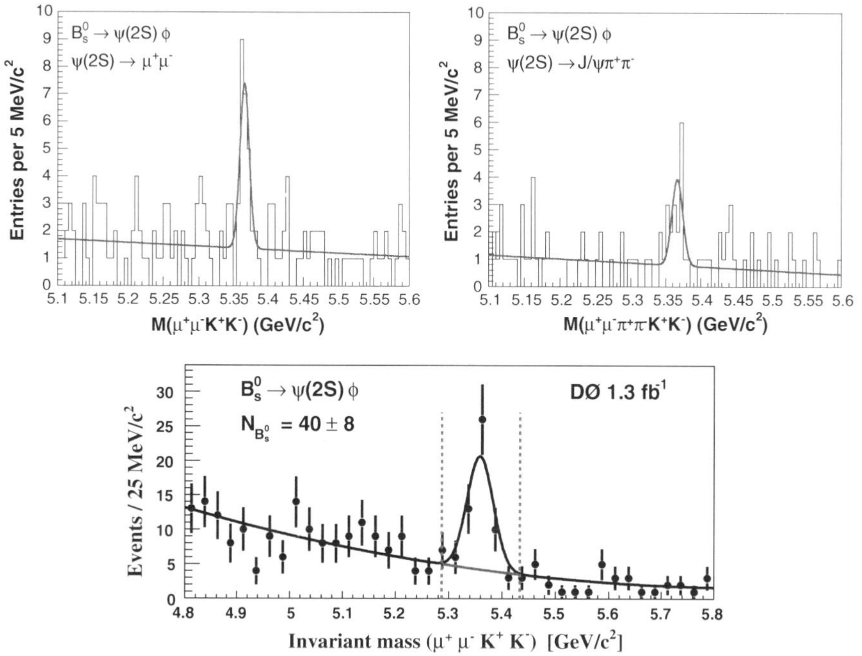

Fig. 5.12 Fits to invariant mass spectra of $B_s^0 \to \psi(2S)\phi$ candidates measured by CDF [17] (*top*) and D0 [18] (*bottom*).

a cut-based method is used for the $B_s^0 \to J/\psi K^{*0}$ mode, the lower signal-to-background ratio in the $B_s^0 \to J/\psi K_S^0$ mode, as expected from the corresponding B^0 decays, calls for a multivariate analysis technique. In this case an artificial neural network is trained with input variables that are independent of the B_s^0 candidate mass and thus do not bias the distribution from which the signal yield is determined. Fit projections to the invariant mass are shown in Fig. 5.13.

The branching ratios are measured relative to the Cabibbo-favored B^0 decay to the same final state. The obtained results of $\mathcal{B}(B_s^0 \to J/\psi K^{*0})/\mathcal{B}(B^0 \to J/\psi K^{*0}) = 0.062 \pm 0.009$ (stat) $\pm\, 0.025$ (syst) $\pm\, 0.008$ (frag) and $\mathcal{B}(B_s^0 \to J/\psi K_S^0)/\mathcal{B}(B^0 \to J/\psi K_S^0) = 0.041 \pm 0.007$ (stat) $\pm\, 0.004$ (syst) $\pm\, 0.005$ (frag), where the last error comes from the uncertainty on f_s/f_d, are in good agreement with the expectation from the spectator model of $\sim |V_{cd}/V_{cs}|^2 \approx \lambda^2 \approx 5\,\%$. In both modes the combinatorial background contributes the largest systematic uncertainty. In the $J/\psi K^{*0}$ mode a sizable systematic uncertainty also comes from the unknown polarization.

A further promising B_s^0 mode for *CP* violation studies is the decay to the *CP* eigenstate $J/\psi f_0(980)$ with $f_0(980) \to \pi^+\pi^-$. This decay was first observed in 2011 by LHCb [21] and briefly afterwards confirmed by Belle [22], CDF [23], and D0 [24]. While the signal yield of the above mentioned $B_s^0 \to J/\psi K_S^0$ decay is still low, the larger yield of $B_s^0 \to J/\psi f_0(980)$ events already enabled CDF to measure the lifetime for this *CP*-odd final state as described in Sect. 5.4.

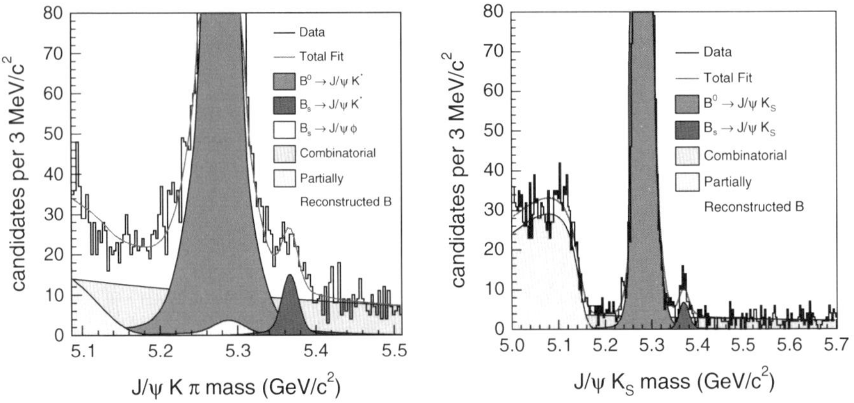

Fig. 5.13 Invariant mass distributions and fit projections for $B_s^0 \to J/\psi K^{*0}$ (*left*) and $B_s^0 \to J/\psi K_S^0$ candidates (*right*) [19]

Both experiments, CDF and D0, measure the $B_s^0 \to J/\psi f_0(980)$ branching ratio relative to the $B_s^0 \to J/\psi\phi$ mode:

$$R_{f_0/\phi} = \frac{\mathcal{B}(B_s^0 \to J/\psi f_0(980))}{\mathcal{B}(B_s^0 \to J/\psi\phi)} \frac{\mathcal{B}(f_0 \to \pi^+\pi^-)}{\mathcal{B}(\phi \to K^+K^-)}. \tag{5.6}$$

The yields of signal and normalization modes are obtained from fits to B_s^0 candidates invariant mass distributions. D0 corrects the $B_s^0 \to J/\psi\phi$ yield for a $(12 \pm 3)\%$ $B_s^0 \to J/\psi K^+K^-$ contribution with a K^+K^- S-wave as observed in their dedicated analysis of this decay [25] (see Chap. 7). As all CDF measurements are consistent with no S-wave contribution, they do not apply such a correction. The measured relative branching ratios of $R_{f_0/\phi} = 0.257 \pm 0.020\,(\text{stat}) \pm 0.014\,(\text{syst})$ for CDF [23] and $R_{f_0/\phi} = 0.275 \pm 0.041\,(\text{stat}) \pm 0.061\,(\text{syst})$ for D0 [24] are compared with other measurements in Fig. 5.14. The main source of systematic uncertainties is the background model in both cases. In the CDF analysis also the uncertainties of the $f_0(980)$ decay model parameters become important.

The decay $B_s^0 \to \phi\phi$ is a penguin-dominated $b \to s\bar{s}s$ transition and therefore considered sensitive to new physics effects. First evidence for this decay was seen by CDF in a data sample with an integrated luminosity of 180 pb^{-1} [26]. Using a signal with a significance of $4.7\,\sigma$ the branching ratio relative to $B_s^0 \to J/\psi\phi$ was measured. This measurement was updated by CDF to $\mathcal{B}(B_s^0 \to \phi\phi)/\mathcal{B}(B_s^0 \to J/\psi\phi) = [1.78 \pm 0.14\,(\text{stat}) \pm 0.20\,(\text{syst})] \times 10^{-2}$ using a data sample corresponding to an integrated luminosity of 2.9 fb^{-1} [27]. The invariant mass spectra used in the two measurements are shown in Fig. 5.15. The increased signal yield of approximately 300 events allows to study this decay in more detail.

Because the daughter particles of the pseudoscalar B_s^0 meson are vector mesons, like in the decay $B_s^0 \to J/\psi\phi$ discussed in detail in Sect. 7.1, the decay is governed

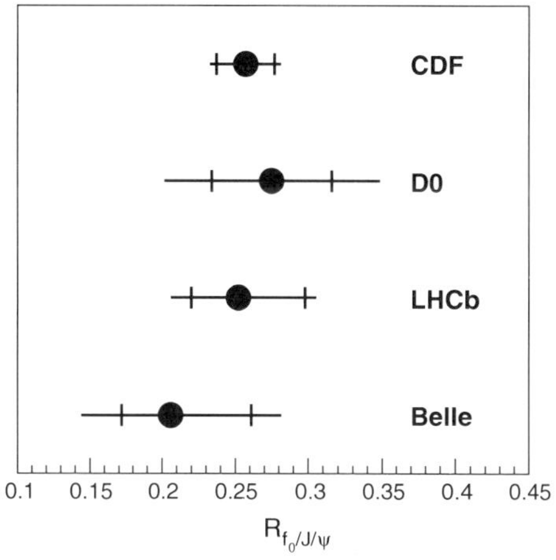

Fig. 5.14 Comparison of measured $R_{f_0/\phi}$ values [21–24], where the Belle result is calculated from an absolute branching ratio measurement

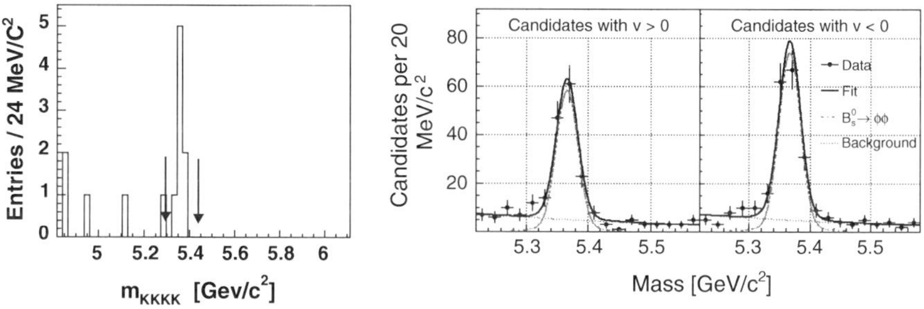

Fig. 5.15 Invariant mass spectra of $B_s^0 \to \phi\phi$ candidates in 180 pb^{-1} (*left*) and 2.9 fb^{-1} of CDF data (*right*) [26, 27]. In the latter plot the data is split into two bins in $v = \sin\Phi$, where Φ is the angle between the ϕ decay planes in the B_s^0 rest frame

by three independent amplitudes, A_0, $A_{||}$, and $A_{\perp}$, corresponding to longitudinal polarization and transverse polarization with parallel and perpendicular spins. The first two states are *CP*-even, the last one is *CP*-odd. The polarization amplitudes can be measured from the distribution of three decay angles, defined in the helicity basis in this analysis. The differential decay rate as a function of these angles can be separated in six terms, three containing squared amplitudes, and three interference terms containing the product of two different amplitudes. In case of negligible *CP* violation, as predicted in the standard model, the two interference terms of *CP*-even and *CP*-odd amplitudes vanish.

This angular function is used in a fit to the data to determine the polarization amplitudes. The reconstructed invariant mass is used as further observable in the fit to separate signal and background. The measured longitudinal fraction of $f_L = |A_0|^2/(|A_0|^2 + |A_{||}|^2 + |A_{\perp}|^2) = 0.348 \pm 0.041\ (\text{stat}) \pm 0.021\ (\text{syst})$ [27] is smaller than in other B meson decays to two vector mesons with $b \to s$ transitions [28, 29]. Because of the $V-A$ structure of the weak interaction one would expect the longitudinal polarization to dominate. The result suggests that penguin annihilation processes have a large contribution [30].

In case of new physics contributions, the two interference terms between CP-even and CP-odd amplitudes do not vanish in the differential decay rate any more in general. To check for such contributions, the following asymmetries are determined

$$A_{u(v)} = \frac{N^+_{u(v)} - N^-_{u(v)}}{N^+_{u(v)} + N^-_{u(v)}} = \mathcal{N}_{u(v)}[\mathrm{Im}(A^*_{||(0)}A_\perp) + \mathrm{Im}(\bar{A}^*_{||(0)}\bar{A}_\perp)], \tag{5.7}$$

where $N^\pm_{u(v)}$ are the number of signal events with positive and negative value of u (v), respectively, and $\mathcal{N}_{u(v)}$ is a normalization factor. The variables u and v are defined as $\sin 2\Phi$ and $\sin\Phi$, respectively, where Φ is the angle between the two ϕ meson decay planes in the B^0_s rest frame. The results of $A_u = -0.007 \pm 0.064$ (stat)± 0.018 (syst) and $A_v = -0.120 \pm 0.064$ (stat) ± 0.016 (syst) [27] are consistent with the standard model prediction of no asymmetry. With future higher statistics samples of this decay, new physics effects may be searched for in a time dependent angular analysis which would complement such measurements in $B^0_s \to J/\psi\phi$ decays that are presented in Chap. 7.

5.3 $B^0_s \to D^{(*)+}_s D^{(*)-}_s$

One of the features that make the B^0_s system distinct from the B^0 system is the sizable decay width difference between the mass eigenstates, $\Delta\Gamma_s = \Gamma(B^0_{sL}) - \Gamma(B^0_{sH})$, which leads to an increased sensitivity to CP-violating effects as discussed in Chap. 7. Because of the relevance of the $B^0_s \to D^{(*)+}_s D^{(*)-}_s$ decay mode for $\Delta\Gamma_s$, as explained below, it is discussed separately from the other B^0_s decay modes. In the standard model, $\Delta\Gamma_s$ is expected to be 0.087 ± 0.021 ps [31, 32] and thus of the order of several percent of the mean decay width, Γ_s. The reason for this non-negligible $\Delta\Gamma_s$ value are the B^0_s decays to CP-even final states that are, in case of negligible direct CP violation, only accessible by the CP-even B^0_s eigenstate. The additional decays modes of the CP-even B^0_s eigenstate lead to a higher total decay rate of this state and thus to a lower lifetime. The decay width difference between the CP eigenstates, $\Delta\Gamma^{CP}_s = \Gamma(B^{CP+}_s) - \Gamma(B^{CP-}_s)$, is related to the off-diagonal element of the decay matrix by $\Delta\Gamma^{CP}_s = 2|\Gamma_{12}|$. According to Eq. (3.18) one can therefore express $\Delta\Gamma_s$ as:

$$\Delta\Gamma_s = \Delta\Gamma^{CP}_s \cos\phi \tag{5.8}$$

In the standard model ϕ is to good approximation zero so that $\Delta\Gamma_s$ is given directly by the difference between the decay rates of the CP-even and CP-odd B^0_s eigenstates.

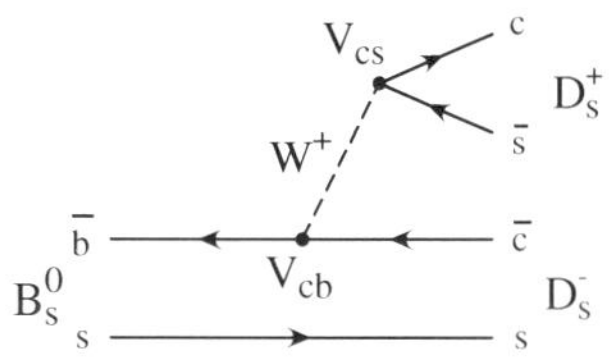

Fig. 5.16 Leading order Feynman diagram of the $B_s^0 \to D_s^{(*)+} D_s^{(*)-}$ decay

One way to experimentally determine $\Delta\Gamma_s$ is to measure the lifetimes in decays to CP-even and CP-odd final states, like $B_s^0 \to J/\psi f_0(980)$ as described in the next section. The other approach is to measure the rates of decays, that are only accessible by the CP-even or CP-odd eigenstates.

The dominant contribution to $\Delta\Gamma_s$ is believed to come from the $B_s^0 \to D_s^{(*)+} D_s^{(*)-}$ decays as explained in the following. The decays are Cabibbo-favored and not color-suppressed as shown in Fig. 5.16 and therefore have a large branching fraction. The corresponding B^0 decays, $B^0 \to D^{(*)+} D^{(*)-}$, are Cabibbo-suppressed, leading to a decay width difference in the B^0 system that is below the current experimental sensitivity.

The final state in $B_s^0 \to D_s^{(*)+} D_s^{(*)-}$ decays is predominantly CP-even. For the B_s^0 decay to two D_s mesons the orbital angular momentum of the final state, L, is zero so that this final state is a pure CP-even state. But in the case of two D_s^* mesons, L can have the values 0, 1, or 2 so that there is a contribution to $\Gamma(B_s^{CP-})$ from the $L = 1$ final state. This term vanishes in the Shiftman-Voloshin limit [33], where the charm quark mass is half the b quark mass and the number of colors goes to infinity. Under these assumptions, all $B_s^0 \to D_s^{(*)+} D_s^{(*)-}$ decays are CP-even and they saturate $\Delta\Gamma_s$, so that the following relation is obtained [34]:

$$2\mathscr{B}(B_s^0 \to D_s^{(*)+} D_s^{(*)-}) \approx \frac{\Delta\Gamma_s}{\Gamma_s + \Delta\Gamma_s/2}. \tag{5.9}$$

However, recent theoretical calculations suggest that three-body modes may provide a significant contribution to $\Delta\Gamma_s$ [35].

A first measurement of $B_s^0 \to D_s^{(*)+} D_s^{(*)-}$ decays was performed by ALEPH using correlations between two ϕ mesons from $D_s \to \phi\pi$ decays [36]. A more exclusive reconstruction method was employed by D0 on a data sample with an integrated luminosity of 1.3 fb^{-1} [37] and, in an updated analysis, of 2.8 fb^{-1} [38]. One of the D_s daughter particles is reconstructed in the decay mode $D_s^+ \to \phi\pi^+$ with $\phi \to K^+K^-$. The other D_s daughter particle is reconstructed partially in the semileptonic decay $D_s^- \to \phi\mu^-\bar{\nu}$. These events are mainly selected online by the single muon trigger. No attempt is made to reconstruct the photon or neutral pion from a D_s^* decay. So the $B_s^0 \to D_s^{(*)+} D_s^{(*)-}$ modes are not separated and all of them contribute to the signal.

The number of events with two correlated D_s mesons is determined in a two-dimensional fit to the invariant mass of the fully reconstructed D_s meson

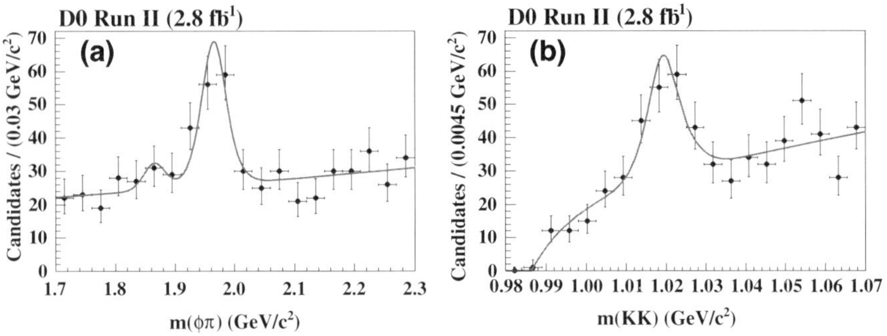

Fig. 5.17 Invariant mass distribution of $D_s^+ \to \phi\pi^+$ (*left*) and $\phi \to K^+K^-$ candidates from semileptonic D_s decays (*right*) with projections of a two-dimensional fit [38]. The fit includes a component for $D^+ \to \phi\pi^+$ decays resulting in the peak at $m(\phi\pi) = 1.87$ GeV/c^2

candidate and the invariant mass of the ϕ from the partially reconstructed D_s meson candidate. Fit projections are shown in Fig. 5.17. The yield of 31.0 ± 9.4 events still contains background contributions from D_s pairs produced in $p\bar{p} \to c\bar{c}X$ events, $B_s^0 \to D_s^{(*)}\phi\mu^-\bar{\nu}$ decays, and $B^{0,\pm} \to D_s^{(*)}D_s^{(*)}KX$ decays. These backgrounds are estimated from mass sidebands and simulation. After subtracting their contribution a signal yield of 26.6 ± 8.4 events is obtained. The significance of the signal corresponds to 3.2 standard deviations.

For the branching ratio measurement, the yield is normalized to the yield of $B_s^0 \to D_s^{(*)}\phi\mu^-\bar{\nu}$ events. Using a relative efficiency estimation from simulation and world average D_s branching ratios, an inclusive branching ratio of $\mathscr{B}(B_s^0 \to D_s^{(*)+}D_s^{(*)-}) = (3.5 \pm 1.0\,(\text{stat}) \pm 0.8\,(\text{syst}) \pm 0.7\,(\mathscr{B}))\,\%$ is measured [38], where the last uncertainty stems from the external inputs of the branching fractions. The dominant experimental systematic uncertainty is the uncertainty of the relative reconstruction efficiency.

The exclusive decay $B_s^0 \to D_s^+D_s^-$ was first observed by CDF using events selected by the Two-Track-Trigger in 355 pb^{-1} of data [39]. One D_s meson is reconstructed in the decays to $\phi\pi^+$ with $\phi \to K^+K^-$, $\bar{K}^{*0}K^+$ with $\bar{K}^{*0} \to K^+\pi^-$, or $\pi^+\pi^-\pi^+$, and the other in the mode $D_s^- \to \phi\pi^-$. Because of the expected low signal-to-background ratio no combinations are considered where the second D_s decay to $K^{*0}K^-$ or $\pi^-\pi^+\pi^-$. In the $B_s^0 \to D_s^+(\bar{K}^{*0}K^+)D_s^-(\phi\pi^-)$ mode, misreconstructed $B^0 \to D^+(K^+\pi^-\pi^+)D_s^-(\phi\pi^-)$ decays, where the π^- from the D^+ is assigned the kaon mass hypothesis, lead to a reflection that peaks close to the B_s^0 signal. This background is removed by vetoing events where the selected final state particles are consistent with the B^0 decay hypothesis.

The full reconstruction of the $B_s^0 \to D_s^+D_s^-$ decay allows to extract the signal yield from a fit to the B_s^0 candidate invariant mass spectrum. This fit is limited to masses above 5.3 GeV/c^2 to exclude $B_s^0 \to D_s^{*+}D_s^{(*)-}$ and other partially reconstructed B meson decays. As normalization channel the kinematically very similar $B^0 \to D_s^+D^-$ decay with $D^- \to K^+\pi^-\pi^-$ and the same three D_s^+ decay modes is

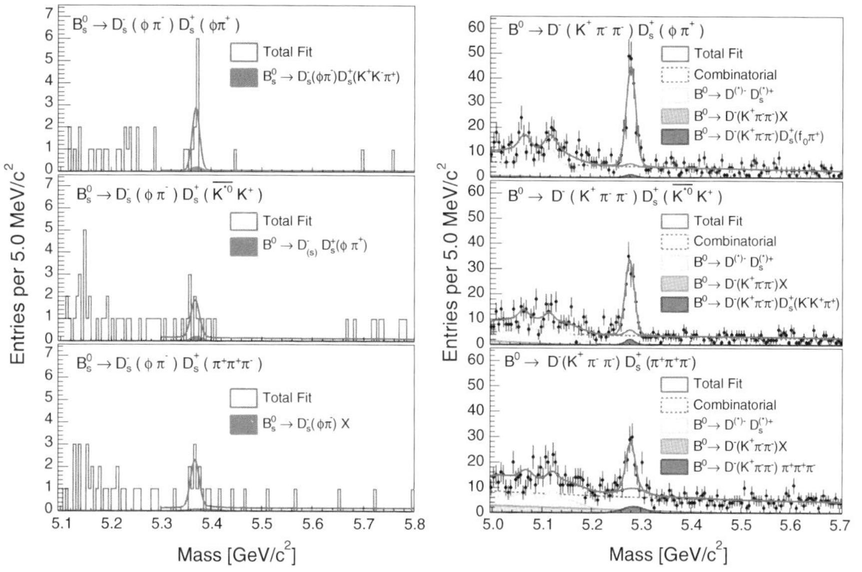

Fig. 5.18 Invariant mass distributions of $B_s^0 \to D_s^+ D_s^-$ (*left*) and $B^0 \to D_s^+ D^-$ candidates (*right*) with fit projections [39]

used. Fit projections are shown in Fig. 5.18. The significance of the signal combined over all three channels is 7.5 standard deviations.

The measured relative branching fraction is $\mathscr{B}(B_s^0 \to D_s^+ D_s^-)/\mathscr{B}(B^0 \to D_s^+ D^-) = 1.44^{+0.38}_{-0.31}\,(\text{stat})^{+0.08}_{-0.12} \pm 0.21\,(\text{frag}) \pm 0.02\,(\mathscr{B})$, where the last two uncertainties are due to the uncertainties of f_s/f_d and the $D_s^+ \to \phi\pi^+$ and $D^+ \to K^+\pi^-\pi+$ branching ratios. For the absolute branching fraction a value of $\mathscr{B}(B_s^0 \to D_s^+ D_s^-) = (0.94^{+0.44}_{-0.42})\,\%$ is derived [39].

In an updated analysis of 6.8 fb^{-1} of data, CDF was able to also observe the $B_s^0 \to D_s^{*\pm} D_s^{\mp}$ and $B_s^0 \to D_s^{*+} D_s^{*-}$ decays [40]. The same B_s^0 and B^0 final states as in the previous analysis are used, except for the mode with $D_s^+ \to \pi^+\pi^-\pi^+$ because of its low signal significance. The selection is optimized with a neural network that uses input variables common to the B_s^0 signal and B^0 normalization mode to minimize systematic uncertainties.

The photon and the neutral pion from the $D_s^{*+} \to D_s^+\gamma$ and $D_s^{*+} \to D_s^+\pi^0$ decays are not reconstructed because of their low detection efficiency. Thus the partially reconstructed $B_s^0 \to D_s^{*\pm} D_s^{\mp}$ and $B_s^0 \to D_s^{*+} D_s^{*-}$ decays show up as broad peaks at values below the nominal B_s^0 mass in the reconstructed $B_s^0 \to D_s^+ D_s^-$ invariant mass distribution. The reflections of $B^0 \to D^+(K^+\pi^-\pi^+)D_s^-(\phi\pi^-)$ reconstructed as $B_s^0 \to D_s^+(\bar{K}^{*0}K^+)D_s^-(\phi\pi^-)$, that were vetoed in the previous analysis, are here taken into account by constraining the relative yields of B^0 signal and reflections in a simultaneous fit. In the simultaneous fit of the two signal and two normalization modes, the shapes of the partially reconstructed and reflection

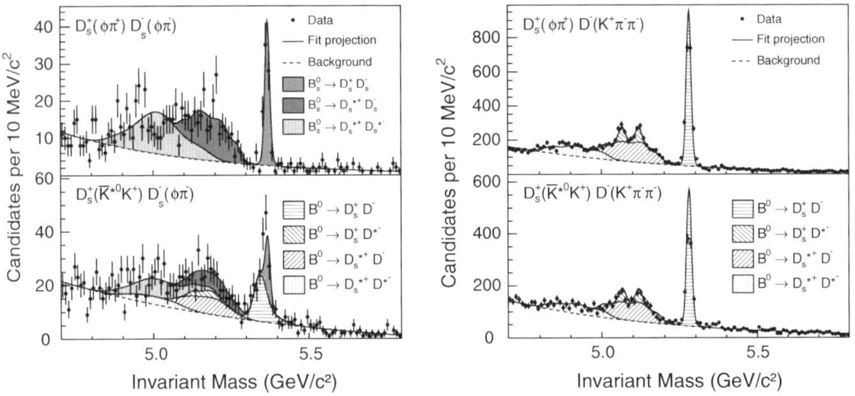

Fig. 5.19 Invariant mass distributions of $B_s^0 \to D_s^+ D_s^-$ (*left*) and $B^0 \to D_s^+ D^-$ candidates (*right*) with fit projections [40]

Table 5.1 Measured relative and absolute $B_s^0 \to D_s^{(*)+} D_s^{(*)-}$ branching ratios [40].

Mode X	$f_s \mathscr{B}(B_s^0 \to X)/f_d \mathscr{B}(B^0 \to D^+ D_s^-)$	$\mathscr{B}(B_s^0 \to X)$ [%]
$B_s^0 \to D_s^+ D_s^-$	$0.183 \pm 0.021 \pm 0.017$	$0.49 \pm 0.06 \pm 0.05 \pm 0.08$
$B_s^0 \to D_s^{*\pm} D_s^{\mp}$	$0.424 \pm 0.046 \pm 0.035$	$1.13 \pm 0.12 \pm 0.09 \pm 0.19$
$B_s^0 \to D_s^{*+} D_s^{*-}$	$0.654 \pm 0.072 \pm 0.065$	$1.75 \pm 0.19 \pm 0.17 \pm 0.29$
$B_s^0 \to D_s^{(*)+} D_s^{(*)-}$	$1.261 \pm 0.095 \pm 0.112$	$3.38 \pm 0.25 \pm 0.30 \pm 0.56$

The uncertainties are statistical, systematic, and due to the normalization

components are taken from simulation. The fit projection is shown in Fig. 5.19. The significance of each of the three B_s^0 decays exceeds 10 standard deviations.

Contributions to the reconstructed D_s decay modes from non-resonant $D_s^+ \to K^- \pi^+ \pi^+$ decays, decays to other resonances, and interference effects are taken into account in the acceptance calculation by simulating events according to the D_s Dalitz structure measured by CLEO [41]. The uncertainty of this Dalitz structure contributes a 6% uncertainty to the B_s^0 branching fractions. A systematic error contribution of similar size comes from the D_s^+ and D^+ branching ratios and, in case of the $B_s^0 \to D_s^{*+} D_s^{*-}$ modes, from the background model. For the absolute branching fractions the dominant systematic error is the uncertainty on f_s/f_d and the branching ratio of the $B^0 \to D^+ D_s^-$ normalization mode. The measured relative and absolute branching ratios are given in Table 5.1

Assuming Eq. (5.9) to hold, the measured inclusive branching ratios can be used to determine $\Delta\Gamma_s/\Gamma_s$. A comparison of the Tevatron results with each other, results from ALEPH and Belle, and the theoretical prediction is shown in Fig. 5.20. The fact that the CDF and D0 measurements are both below the predicted value may be a hint for a sizable contribution of three-body decay modes to $\Delta\Gamma_s$.

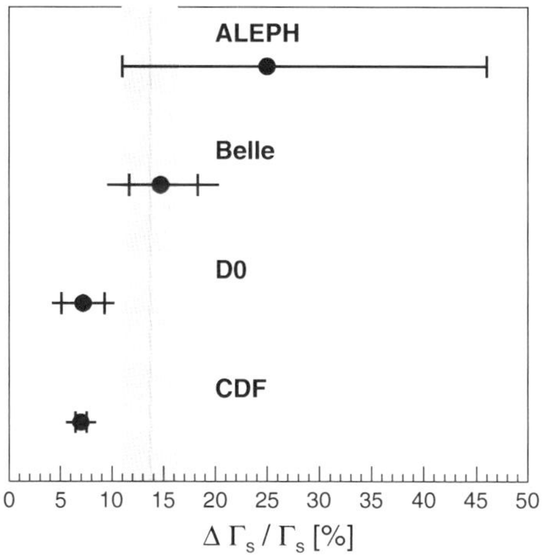

Fig. 5.20 Comparison of $\Delta\Gamma_s/\Gamma_s$ values derived from $B_s^0 \to D_s^{(*)+} D_s^{(*)-}$ branching ratio measurements [36, 38, 40, 42] using Eq. (5.9) and the theoretical prediction shown as vertical band [32]

5.4 B_s^0 Lifetime

Compared to lifetime measurements of B^+ and B^0 mesons, where the decay rate distribution follows an exponential function, the sizable decay width difference makes the situation for B_s^0 lifetime measurements more complicated. For an untagged measurement, meaning that the flavor of the B_s^0 meson at production is not identified, we have to consider the sum of B_s^0 and $\bar{B}_s^0$ decays to the same final state that are given in Eqs. (3.38) and (3.39). Assuming no *CP* violation in mixing ($|p/q| = 1$), the cos and sin terms in the sum cancel and one obtains

$$\begin{aligned}
\frac{d\Gamma(t)}{dt} &= \frac{d\Gamma[B \to f](t)}{dt} + \frac{d\Gamma[\bar{B} \to f](t)}{dt} \\
&= N_f |A_f|^2 e^{-\Gamma t} \left[(1 + |\lambda_f|^2) \cosh \frac{\Delta\Gamma}{2} t - 2\mathrm{Re}(\lambda_f) \sinh \frac{\Delta\Gamma}{2} t \right] \\
&= N_f |A_f|^2 e^{-\Gamma t} \left[\frac{1 - 2\mathrm{Re}(\lambda_f) + |\lambda_f|^2}{2} e^{\frac{\Delta\Gamma}{2} t} + \frac{1 + 2\mathrm{Re}(\lambda_f) + |\lambda_f|^2}{2} e^{-\frac{\Delta\Gamma}{2} t} \right] \\
&= N_- e^{-\left(\Gamma - \frac{\Delta\Gamma}{2}\right) t} + N_+ e^{-\left(\Gamma + \frac{\Delta\Gamma}{2}\right) t}
\end{aligned} \tag{5.10}$$

with

$$N_\pm = N_f |A_f|^2 \frac{1 \pm 2\mathrm{Re}(\lambda_f) + |\lambda_f|^2}{2}. \tag{5.11}$$

So the experimentally observed decay rate distribution is the sum of two exponential functions. The normalizations of the long and short lived component, N_- and N_+,

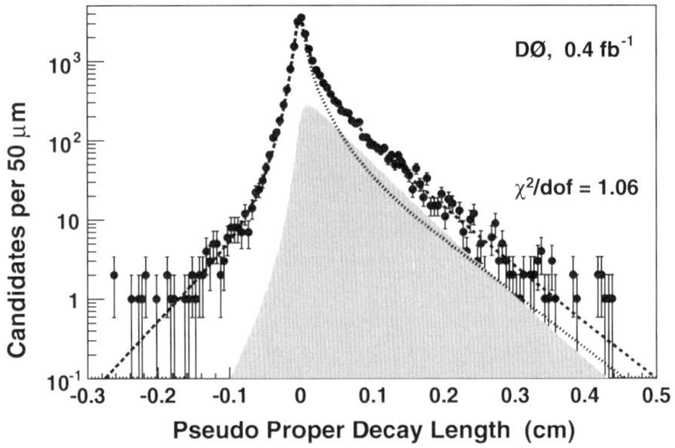

Fig. 5.21 Visible proper decay length distribution of $B_s^0 \to D_s^- \mu^+ \nu X$ candidates with fit projection [43]. The filled area represents the the signal, the *dashed line* the combinatorial background, and the *dotted line* the sum of both components

respectively, depend on the final state. In the case of a flavor specific final state, like for semileptonic decays, λ_f is zero and both normalizations become identical. Other special and interesting cases are final states where one of the terms vanishes so that the decay time distribution is again a single exponential. In the standard model, where the *CP* violation in the B_s^0 system is negligible, this happens for *CP* eigenstates where $\lambda_f = \pm 1$.

Like in B^+ and B^0 lifetime measurements, different techniques are also employed for B_s^0 lifetime measurements. Using 400 pb^{-1} of data, D0 has measured an effective lifetime of B_s^0 mesons in semileptonic decays [43]. The B_s^0 mesons are reconstructed via $B_s^0 \to D_s^- \mu^+ X$ with $D_s^- \to \phi\pi^-$ and $\phi \to K^+K^-$. The distribution of the visible proper decay time as defined in Eq. (5.3) is fitted to determine the B_s^0 lifetime. In this measurement the effect of $\Delta\Gamma_s$ is neglected and a single exponential function convolved with a Gaussian resolution function and a K factor distribution to account for the incomplete B_s^0 reconstruction is used for the signal. Wrong-sign $D_s^- \mu^-$ events and D_s^- mass sideband events are used to constrain the t_{vis} distribution of combinatorial background events. Further background contributions are from other B meson decays and $c\bar{c}$ events. A projection of the fit is shown in Fig. 5.21.

The measured effective lifetime in the flavor-specific semileptonic decays is $\tau_{fs}(B_s^0) = [1.398 \pm 0.044\,(\text{stat})^{+0.028}_{-0.025}\,(\text{syst})]$ ps [43]. The largest contribution to the systematic uncertainty comes from the uncertainties on the backgrounds.

A measurement of the effective B_s^0 lifetime in flavor-specific hadronic decays was performed by CDF using events selected by the Two-Track-Trigger in a data sample of 1.3 fb^{-1} [44]. The reconstructed decay is $B_s^0 \to D_s^- \pi^+$ with $D_s^- \to \phi\pi^-$ and $\phi \to K^+K^-$. The invariant mass distribution of the selected B_s^0 candidates is shown in Fig. 5.22 (left). One can see a peak at the nominal B_s^0 mass from correctly reconstructed $B_s^0 \to D_s^- \pi^+$ decays. The decay time of these fully reconstructed (FR) decays is directly obtained from Eq. (5.2). The invariant mass region below

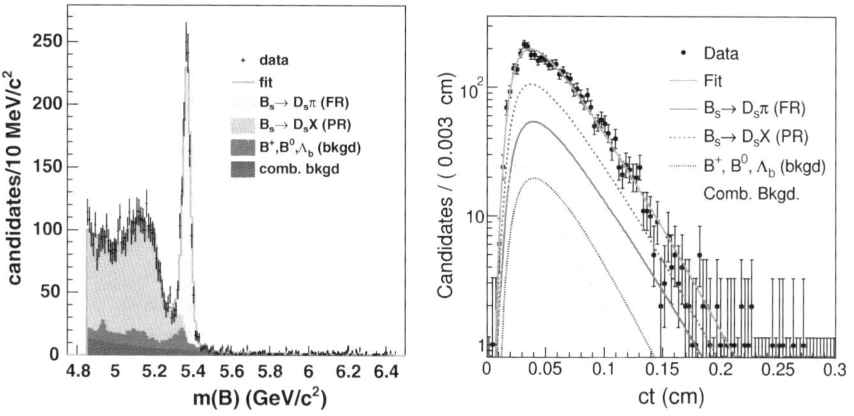

Fig. 5.22 Invariant mass (*left*) and decay time distribution (*right*) of $B_s^0 \to D_s^- \pi^+$ candidates with fit projection [44]

the nominal B_s^0 mass is populated by partially reconstructed (PR) b hadron decays, including $B_s^0 \to D_s^- \pi^+ (n\gamma)$, $B_s^0 \to D_s^{\pm} K^{\mp}$, $B_s^0 \to D_s^- \rho^+$, $B_s^0 \to D_s^{*-} \pi^+$, and other $B_s^0 \to D_s^{(*)-} X$ decays. These B_s^0 decays are included in the lifetime measurement to increase the sensitivity. Although a K factor has to be used, the correction is small because of the selected invariant mass range. And the usage of the reconstructed instead of the nominal mass in Eq. (5.3) reduced the width of the K factor distribution to a few percent so that the partially reconstructed hadronic decays provide a similar sensitivity to the B_s^0 lifetime as the fully reconstructed events.

The yields of the fully and all partially reconstructed B_s^0 and other b hadron modes are determined in a fit to the invariant mass distribution. The shapes of each component are obtained from simulation. The backgrounds from real and fake D_s^- mesons combined with a random track are taken from wrong-sign events and D_s^- mass sidebands, respectively.

The fractions of components determined in the mass fit are then used in a fit of the decay time. The fully and partially reconstructed signal PDFs take the time dependent trigger acceptance into account by a multiplicative efficiency function derived from simulation. The trigger efficiency simulation was checked with $J/\psi \to \mu^+ \mu^-$ events collected by the dimuon trigger. The background PDFs are taken from simulation or mass sidebands. The result of the fit is shown in Fig. 5.22 (right).

Assuming a single exponential decay time distribution for the signal, an effective lifetime of $\tau_{fs}(B_s^0) = [1.518 \pm 0.041 \text{ (stat)} \pm 0.027 \text{ (syst)}]$ ps is measured [44]. The main systematic uncertainty come from the background shapes and fractions, followed by the uncertainty on the trigger simulation. Although this is a measurement for a flavor-specific decay, the two B_s^0 eigenstates do not contribute with the same weight as discussed above. This effect is caused by the time dependent trigger acceptance. The correction factor for this bias depends on $\Delta\Gamma_s$ and is estimated to be $\delta\tau = -0.11(\Delta\Gamma_s/\Gamma_s)^2$.

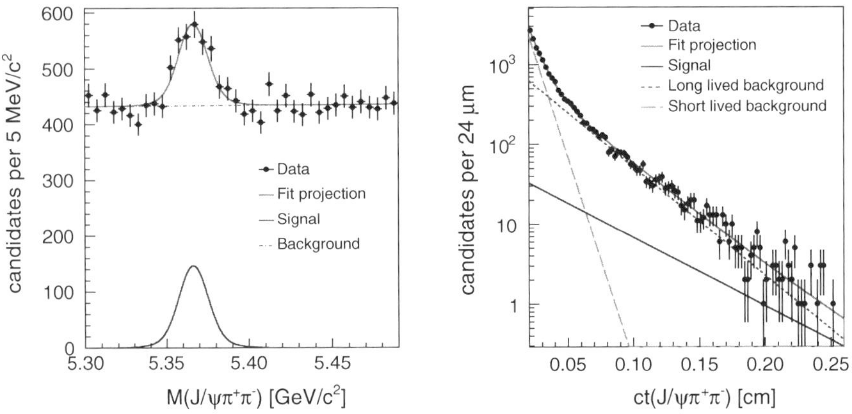

Fig. 5.23 Fit to the invariant mass (*left*) and decay time distribution (*right*) of $B_s^0 \to J/\psi f_0(980)$ candidates [23]

Such issues caused by the mixture of B_{sL}^0 and B_{sH}^0 decays does not exists for decay modes that are only accessible by one of the eigenstates. In the standard model these are the decays to *CP* eigenstates, like $B_s^0 \to J/\psi f_0(980)$ with $f_0(980) \to \pi^+\pi^-$. The lifetime in this mode was first measured by CDF using 3.8 fb^{-1} of data [23].

The B_s^0 lifetime is determined in a combined mass, decay time, and decay time uncertainty fit. Fit projections are shown in Fig. 5.23. Events with a decay time below 0.67 ps are excluded from the fit to avoid that the sample is dominated by prompt events. The resulting mean lifetime of the *CP*-odd B_s^0 eigenstate is $\tau_{CP-}(B_s^0) = [1.70^{+0.12}_{-0.11}\,(\text{stat})\pm 0.03\,(\text{syst})]$ ps [23]. Because of the low signal-to-background ratio the largest systematic uncertainty comes from the background decay time model.

In the standard model the result corresponds to the lifetime of the B_{sH}^0 eigenstate. Combined with other lifetime measurements in flavor-specific or *CP*-even modes, it provides information on the decay width difference $\Delta\Gamma_s$.

Finally $B_s^0 \to J/\psi\phi$ decays are used by CDF and D0 for B_s^0 lifetime measurements. Since the final state is a mixture of *CP*-even and *CP*-odd eigenstates that can be distinguished by an angular analysis, this mode provides information about the lifetime of both B_s^0 eigenstates, B_{sH}^0 and B_{sL}^0. The mode is also the one with the best experimental sensitivity to *CP* violation in the B_s^0 system and therefore discussed in detail in Chap. 7.

Figure 5.24 compares the measurements discussed in this section. It includes the latest results of the average B_s^0 lifetime, $\bar{\tau}(B_s^0) = 1/\Gamma_s = 2/[\Gamma(B_{sL}^0) + \Gamma(B_{sH}^0)]$, measured in $B_s^0 \to J/\psi\phi$ decays by CDF and D0. One can see that the lifetime for the $J/\psi f_0(980)$ final state has a higher value than the other measurements, as expected in case of a sizable decay width difference and negligible *CP* violation. Furthermore the measurements of τ_{fs} and $\bar{\tau}$ support the theoretical prediction that the mean B_s^0 lifetime is equal to the B^0 lifetime within 1 % uncertainty [4].

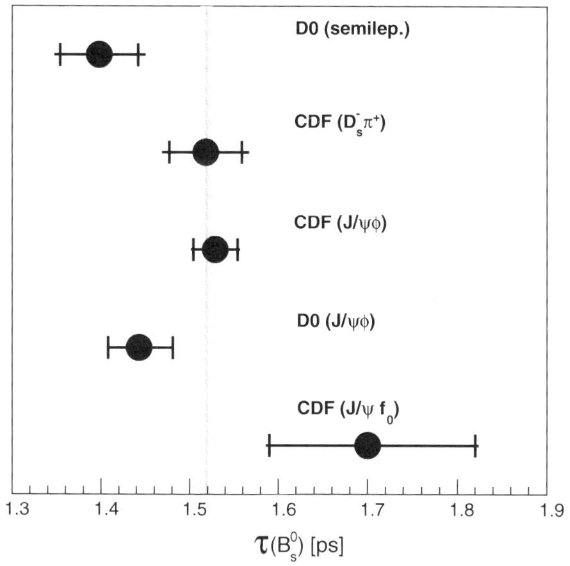

Fig. 5.24 Comparison of B_s^0 lifetime measurements in different final states [23, 25, 43–45]. The vertical band shows the world average B^0 lifetime [10]

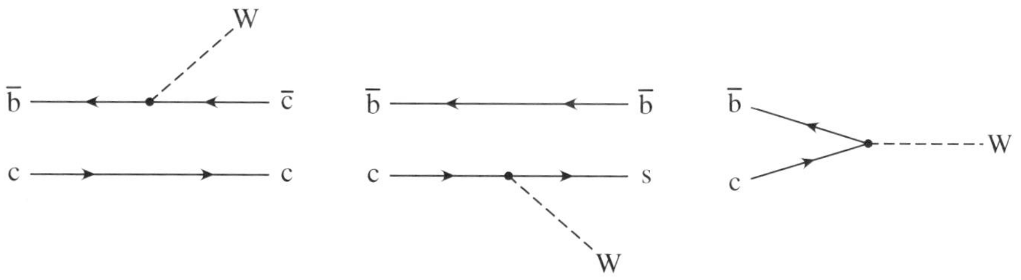

Fig. 5.25 Leading order Feynman diagrams of the three B_c^+ decay processes

5.5 B_c^+

The B_c^+ is a unique particle because it is the only ground state meson with two different heavy quarks. Therefore models for $c\bar{c}$ and $b\bar{b}$ states can be applied to describe the bound state, but it can only decay via the weak interaction. In contrast to the other weakly decaying B mesons, where the lifetime is mainly determined by the decay of the b quark, two further processes contribute to the B_c^+ decay width, namely the decay of the c quark and the annihilation of both heavy quarks as illustrated in Fig. 5.25. These decay processes lead to a reduced lifetime and make the B_c^+ the shortest lived weakly decaying B meson. The B_c^+ lifetime was predicted to be about one third of the B^+ lifetime [46].

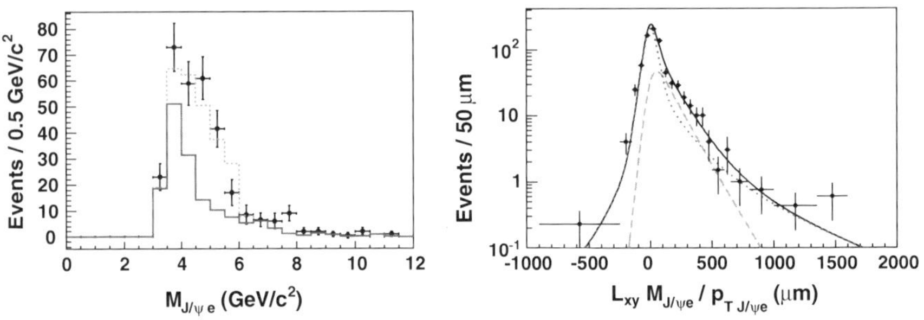

Fig. 5.26 $J/\psi e$ invariant mass distribution for $B_c^+ \to J/\psi e^+ \nu_e$ candidates with $L_{xy}/\sigma(L_{xy}) > 3$ (*left*) and visible proper decay length distribution with fit projection (*right*) [47]. The *solid line* in the *left plot* and the *dotted line* in the *right plot* show the background component

Both Tevatron experiment have measured the B_c^+ lifetime using semileptonic decays. CDF performed the measurement with $B_c^+ \to J/\psi e^+ \nu_e$ events in 360 pb^{-1} of data [47] and D0 with $B_c^+ \to J/\psi \mu^+ X$ events in 1.3 fb^{-1} of data [48]. The J/ψ mesons are reconstructed from muon pairs. In case of CDF, the electrons are selected by requirements on the specific ionization in the drift chamber and the shower shape in the electromagnetic calorimeter.

Several background sources contribute to the selected event samples. The background from random combinations of tracks reconstructed as a J/ψ is estimated from J/ψ mass sidebands. The contribution from real J/ψ combined with a hadron track that is misidentified as a lepton is derived from a sample of J/ψ-track combinations and measured misidentification probabilities. In the CDF measurement, the background from real J/ψ combined with an electron from $\gamma \to e^+e^-$ conversions is estimated from the rate of identified conversions and the efficiency to find conversions. Another type of background are combinations of real J/ψ and real muons, that do not come from a B_c^+ decay. These are mainly $b\bar{b}$ events where the J/ψ and muon stem from two different b hadrons. This background is described with simulated events. Finally, prompt J/ψ in $c\bar{c}$ events combined with a muon track contribute at low decay times. Their fraction is determined in the decay time fit.

Before measuring the lifetime, both experiments establish the existence of a B_c^+ signal by examination of the $J/\psi \ell$ invariant mass spectrum (Figs. 5.26 and 5.27 left). CDF determines the B_c^+ lifetime in a fit of the visible proper decay length and its uncertainty. D0 performs a simultaneous fit of the visible proper decay length and the $J/\psi \mu$ invariant mass. The fit results are shown in the right plots of Figs. 5.26 and 5.27.

CDF measures $\tau(B_c^+) = [0.463^{+0.073}_{0.065}$ (stat) $\pm$ 0.036 (syst)] ps [47], where the dominant uncertainty comes from the background decay time model. The result reported by D0 is $\tau(B_c^+) = [0.448^{+0.038}_{0.036}$ (stat) $\pm$ 0.032 (syst)] ps [48]. Mass and decay time model uncertainties contribute to the systematic uncertainty of the B_c^+ lifetime with about equal amount. Both measurements agree well. In Fig. 5.28 they are compared with theoretical predictions. Although the uncertainties of the

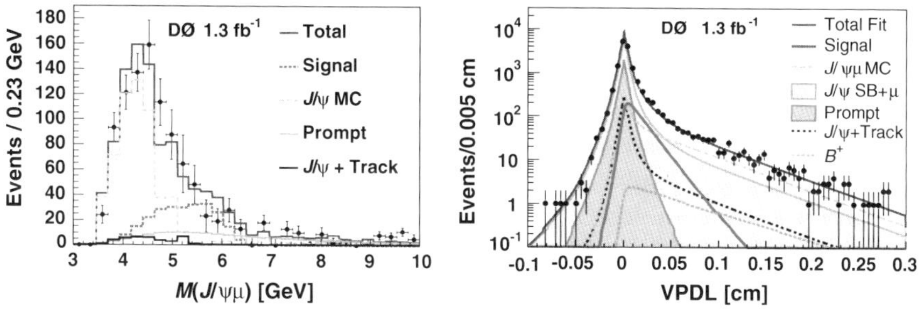

Fig. 5.27 $J/\psi\,\mu$ invariant mass distribution for $B_c^+ \to J/\psi\,\mu^+ X$ candidates with $L_{xy}/\sigma(L_{xy}) > 4$ (*left*) and visible proper decay length (*VPDL*) distribution with fit projection (*right*) [48]

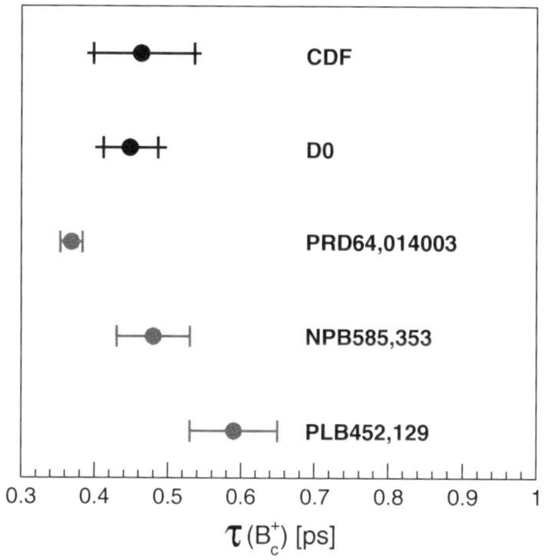

Fig. 5.28 Comparison of B_c^+ lifetime measurements [47, 48] and theoretical predictions [49–51]

measurements and the predictions are still large, the data seem to favor the calculation based on QCD sum rules [49].

5.6 Λ_b^0

CDF and D0 have contributed considerably to the knowledge about b-baryons in general, and Λ_b^0 in particular. Several (relative) Λ_b^0 branching ratios have been measured. The list of decays modes includes $\Lambda_b^0 \to \Lambda_c^+ \mu^- \bar{\nu}_\mu$ [52], $\Lambda_b^0 \to J/\psi\,\Lambda$ [53], $\Lambda_b^0 \to \Lambda_c^+ \pi^-$ [54], $\Lambda_b^0 \to p\pi^-$ and $\Lambda_b^0 \to pK^-$ [54]. The latter three modes have been observed for the first times.

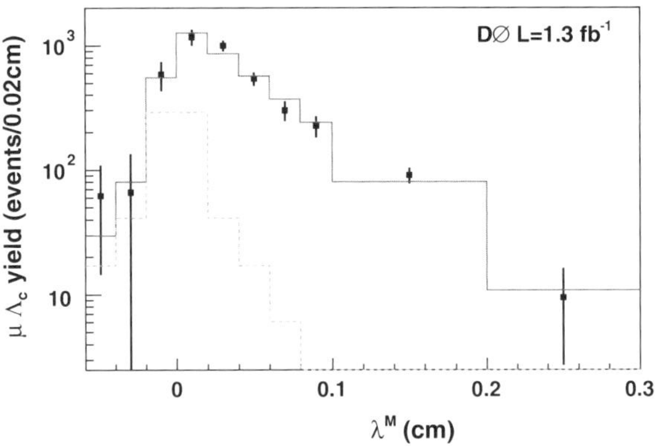

Fig. 5.29 Visible proper decay length distribution of $\Lambda_b^0 \to \Lambda_c^+ \mu^- X$ candidates with fit projection [63]. The *dashed line* shows the background contribution

Of particular interest are Λ_b^0 lifetime measurements because they extend the b hadron lifetime studies to the baryon sector. While first theoretical calculations predicted a lifetime ratio between Λ_b^0 and B^0 in the range 0.9–1 [55–58], measurements at LEP [59–61] and Tevatron Run I [62] yielded a $\tau(\Lambda_b^0)/\tau(B^0)$ value of about 0.8. This discrepancy is known as Λ_b^0 lifetime puzzle. A more recent calculation, that includes next-to-leading order QCD effects and heavy quark expansion terms up to order $1/m_b^4$, has predicted a lifetime ratio of 0.88 ± 0.05 [4] and relaxed the tension between theory and experiment.

Using 1.3 fb^{-1} of data, D0 has measured the Λ_b^0 lifetime in semileptonic $\Lambda_b^0 \to \Lambda_c^+ \mu^- X$ decays [63]. The Λ_c^+ is reconstructed in the decay to pK_S^0 and fitted to a common vertex with a muon candidate. The Λ_c^+ yield is determined in bins of visible proper decay length by a fit to the Λ_c^+ invariant mass distribution. The Λ_b^0 lifetime is then obtained from a fit to this distribution. Background of Λ_c^+ mesons produced in $c\bar{c}$ or $b\bar{b}$ events and combined with a muon from the other heavy quark decay is described using simulation. The signal PDF takes into account contributions from $\Lambda_b^0 \to \Lambda_c^+ \mu^- \bar{\nu}_\mu$ and $\Lambda_b^0 \to \Sigma_c \pi \mu^- \bar{\nu}_\mu$ with $\Sigma_c \to \Lambda_c^+ \pi$.

The result of the fit is shown in Fig. 5.29. The measured lifetime is $\tau(\Lambda_b^0) = [1.290^{+0.119}_{-0.110}\,(\text{stat})^{+0.087}_{-0.091}\,(\text{syst})]$ ps [63], where the largest systematic uncertainty comes from the method to extract the signal yield from the mass fit.

The first measurement of the Λ_b^0 lifetime using fully reconstructed decays was performed by D0 using $\Lambda_b^0 \to J/\psi \Lambda$ with $\Lambda \to p\pi^-$ [64]. This analysis of 250 pb^{-1} of data was later updated with a data sample corresponding to an integrated luminosity of 1.2 fb^{-1} [65]. CDF used the same decay mode to measure the Λ_b^0 lifetime first in a data sample of 1.0 fb^{-1} [66] and then in an updated analysis using 4.3 fb^{-1} [6].

The Λ_b^0 lifetime measurements follow the same basic method as the B^0 and B^+ lifetime measurements in the $J/\psi K$ decay modes described in Sect. 5.1. To suppress $B^0 \to J/\psi K_S^0$ cross contamination, CDF vetos Λ candidates that are con-

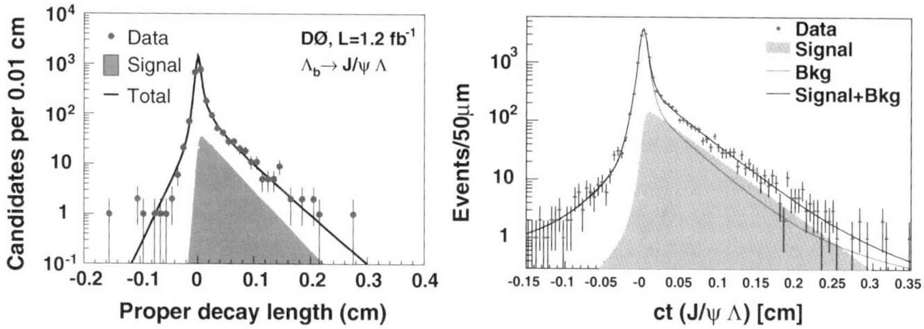

Fig. 5.30 Decay length distributions of $\Lambda_b^0 \to J/\psi\,\Lambda$ candidates measured by D0 (*left*) and CDF (*right*) with fit projections [6, 65]

sistent with $K_S^0 \to \pi^+\pi^-$ decays if the proton candidate is assigned the pion mass hypothesis. D0 suppresses contaminations from cascade decays such as $\Sigma^0 \to \Lambda\gamma$ and $\Xi^0 \to \Lambda\pi^0$ by requiring the Λ flight and momentum direction in the transverse plane to agree within less than 0.8°. The Λ_b^0 lifetime is determined in a fit to the decay time, its uncertainty, the Λ_b^0 candidate invariant mass, and, in case of CDF, the mass uncertainty. Fit projections are shown in Fig. 5.30.

The latest D0 result is $\tau(\Lambda_b^0) = [1.218^{+0.130}_{-0.115}\,(\text{stat}) \pm 0.042\,(\text{syst})]$ ps [65], which is consistent with the earlier D0 measurement of $\tau(\Lambda_b^0) = [1.22^{+0.22}_{-0.18}\,(\text{stat}) \pm 0.04\,(\text{syst})]$ ps [64]. The largest systematic uncertainty comes from the long-lived background component and in particular from the contamination of B^0 decays.

The first CDF measurement in the mode $\Lambda_b^0 \to J/\psi\,\Lambda$ of $\tau(\Lambda_b^0) = [1.593^{+0.083}_{-0.078}\,(\text{stat}) \pm 0.033\,(\text{syst})]$ ps [66] is significantly higher than the previous measurements. The Λ_b^0 lifetime measurement of the updated analysis of $\tau(\Lambda_b^0) = [1.537 \pm 0.045\,(\text{stat}) \pm 0.014\,(\text{syst})]$ ps [6] confirms the result. While the decay time resolution model contributes the largest systematic uncertainty in both analyses, it is reduced in the latter by using the sum of three Gaussian functions instead of just one Gaussian as resolution model.

The Λ_b^0 lifetime was measured in fully hadronic $\Lambda_b^0 \to \Lambda_c^+\pi^-$ decays by CDF using a data sample of 1.1 fb^{-1} [67]. The Λ_c^+ is reconstructed in the decay $\Lambda_c^+ \to pK^-\pi^+$. To suppress background from $D^+ \to K^-\pi^+\pi^+$ decays misreconstructed as Λ_c^+, an upper cut on the reconstructed Λ_c^+ decay length is placed. To determine the composition of the selected sample, the Λ_b^0 invariant mass distribution is fitted. Besides the Λ_b^0 signal, the following background components are included in the fit: combinatorial background, partially and fully reconstructed B meson decays, partially reconstructed Λ_b^0 decays. The shape of the misreconstructed b hadron decay components is obtained from simulation. Fully reconstructed Λ_b^0 decays other than $\Lambda_c^+\pi^+$ are included in the signal. The left plot in Fig. 5.31 shows the fit result. Discrepancies between the fit and the data at low mass values do not affect the Λ_b^0 lifetime result because they are outside the mass window from which events are selected for the following decay time fit.

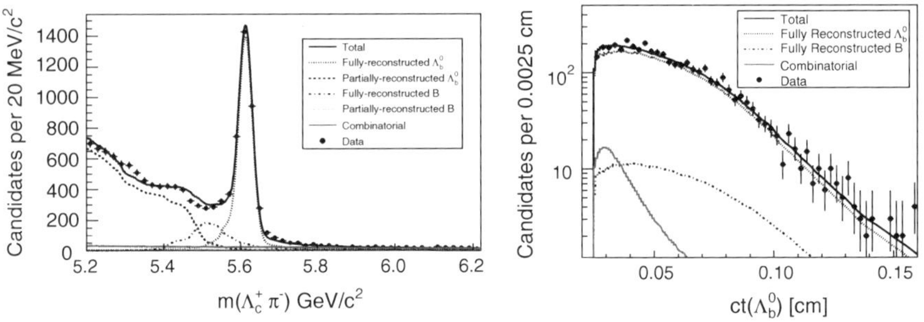

Fig. 5.31 Invariant mass (*left*) and decay length distributions (*right*) of $\Lambda_b^0 \to \Lambda_c^+\pi^+$ candidates with fit projections [67]

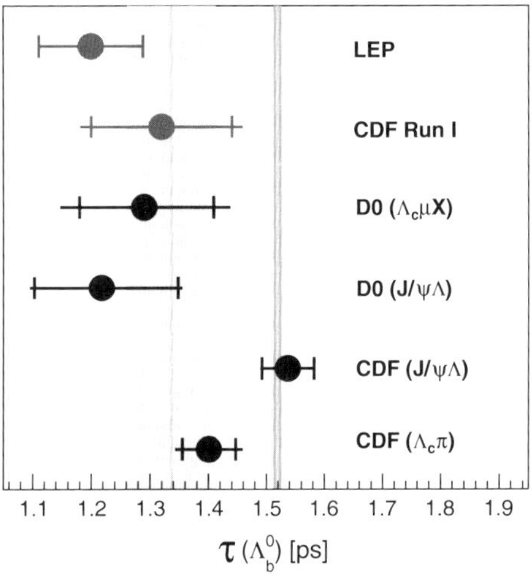

Fig. 5.32 Comparison of Λ_b^0 lifetime measurements from LEP [59–61], Tevatron Run I [62], and Run II [6, 63, 65, 67]. The broad band shows the theoretical expectation, calculated from the predicted $\tau(\Lambda_b^0)/\tau(B^0)$ ratio [4] and the world average B^0 lifetime [10] which is depicted by the narrow band on the right

Since the Λ_b^0 candidates are collected by the Two-Track-Trigger, its time dependent efficiency is determined from simulation and included in the decay time PDF. The reliability of the trigger simulation is checked by a comparison of $J/\psi \to \mu^+\mu^-$ events in data and simulation. The observed discrepancy at the 3-4σ level is assigned as systematic uncertainty, which is the dominant in this measurement. The second largest systematic uncertainty stems from the Λ_c^+ Dalitz model. The measured Λ_b^0 lifetime is $\tau(\Lambda_b^0) = [1.401 \pm 0.046\,(\text{stat}) \pm 0.035\,(\text{syst})]$ ps [67].

Figure 5.32 shows a compilation of Λ_b^0 lifetime measurements. While most measurements are now in good agreement with the theoretical prediction, the most recent and most precise result, the CDF measurement in the $J/\psi\,\Lambda$ decay mode, suggests a higher $\tau(\Lambda_b^0)/\tau(B^0)$ value of the order 1. Whether this is really the case and whether there is a new Λ_b^0 lifetime puzzle with a discrepancy between theory and experiment of opposite sign, has to be clarified in future measurements.

References

1. I.I. Bigi, M.A. Shifman, N. Uraltsev, Ann. Rev. Nucl. Part. Sci. **47**, 591 (1997)
2. M. Beneke, G. Buchalla, C. Greub, A. Lenz, U. Nierste, Nucl. Phys. **B639**, 389 (2002)
3. E. Franco, V. Lubicz, F. Mescia, C. Tarantino, Nucl. Phys. **B633**, 212 (2002)
4. C. Tarantino, Eur. Phys. J. **C33**, S895 (2004)
5. A.J. Lenz, AIP Conf. Proc. **1026**, 36 (2008)
6. T. Aaltonen et al., Phys. Rev. Lett. **106**, 121804 (2011)
7. V.M. Abazov et al., Phys. Rev. Lett. **94**, 182001 (2005)
8. T. Aaltonen et al., Phys. Rev. **D83**, 032008 (2011)
9. G. Punzi, eConf **C030908**, WELT002 (2003).
10. J. Beringer et al., Phys. Rev. D **86**, 010001 (2012)
11. A. Abulencia et al., Phys. Rev. Lett. **96**, 191801 (2006)
12. A. Abulencia et al., Phys. Rev. Lett. **98**, 061802 (2007)
13. W. Yao, et al., J. Phys.G G33, 1 (2006).
14. T. Aaltonen et al., Phys. Rev. Lett. **103**, 191802 (2009)
15. E. Barberio, Z. Was, Comput. Phys. Commun. **79**, 291 (1994)
16. V.M. Abazov et al., Phys. Rev. Lett. **102**, 051801 (2009)
17. A. Abulencia et al., Phys. Rev. Lett. **96**, 231801 (2006)
18. V.M. Abazov et al., Phys. Rev. **D79**, 111102 (2009)
19. T. Aaltonen et al., Phys. Rev. **D83**, 052012 (2011)
20. S. Faller, R. Fleischer, T. Mannel, Phys. Rev. **D79**, 014005 (2009)
21. R. Aaij et al., Phys. Lett. **B698**, 115 (2011)
22. J. Li et al., Phys. Rev. Lett. **106**, 121802 (2011)
23. T. Aaltonen et al., Phys. Rev. **D84**, 052012 (2011)
24. V.M. Abazov et al., Phys. Rev. **D85**, 011103 (2012)
25. V.M. Abazov et al., Phys. Rev. **D85**, 032006 (2012)
26. D.E. Acosta et al., Phys. Rev. Lett. **95**, 031801 (2005)
27. T. Aaltonen et al., Phys. Rev. Lett. **107**, 261802 (2011)
28. B. Aubert et al., Phys. Rev. Lett. **98**, 051801 (2007)
29. K.F. Chen et al., Phys. Rev. Lett. **94**, 221804 (2005)
30. A. Datta, D. London, J. Matias, M. Nagashima, A. Szynkman, Eur. Phys. J. **C60**, 279 (2009)
31. A. Lenz, U. Nierste, JHEP **0706**, 072 (2007)
32. A. Lenz, U. Nierste, TTP11-03, TUM-HEP-792-11 (2011).
33. M.A. Shifman, M. Voloshin, Sov. J.Nucl.Phys. 47, 511 (1988).
34. R. Aleksan, A. Le Yaouanc, L. Oliver, O. Pene, J. Raynal, Phys. Lett. **B316**, 567 (1993)
35. C.K. Chua, W.S. Hou, C.H. Shen, Phys. Rev. **D84**, 074037 (2011)
36. R. Barate et al., Phys. Lett. **B486**, 286 (2000)
37. V.M. Abazov et al., Phys. Rev. Lett. **99**, 241801 (2007)
38. V.M. Abazov et al., Phys. Rev. Lett. **102**, 091801 (2009)
39. T. Aaltonen et al., Phys. Rev. Lett. **100**, 021803 (2008)
40. T. Aaltonen et al., Phys. Rev. Lett. **108**, 201801 (2012)
41. R. Mitchell et al., Phys. Rev. **D79**, 072008 (2009)
42. S. Esen et al., Phys. Rev. Lett. **105**, 201802 (2010)
43. V.M. Abazov et al., Phys. Rev. Lett. **97**, 241801 (2006)
44. T. Aaltonen et al., Phys. Rev. Lett. **107**, 272001 (2011)
45. T. Aaltonen et al., Phys. Rev. **D85**, 072002 (2012)
46. M. Beneke, G. Buchalla, Phys. Rev. **D53**, 4991 (1996)
47. A. Abulencia et al., Phys. Rev. Lett. **97**, 012002 (2006)
48. V.M. Abazov et al., Phys. Rev. Lett. **102**, 092001 (2009)
49. V. Kiselev, A. Kovalsky, A. Likhoded, Nucl. Phys. **B585**, 353 (2000)
50. C.H. Chang, S.L. Chen, T.F. Feng, X.Q. Li, Phys. Rev. **D64**, 014003 (2001)
51. A.Y. Anisimov, I. Narodetsky, C. Semay, B. Silvestre-Brac, Phys. Lett. **B452**, 129 (1999)
52. T. Aaltonen et al., Phys. Rev. **D79**, 032001 (2009)

53. V.M. Abazov et al., Phys. Rev. **D84**, 031102 (2011)
54. A. Abulencia et al., Phys. Rev. Lett. **98**, 122002 (2007)
55. I.I. Bigi, B. Blok, M.A. Shifman, N. Uraltsev, A.I. Vainshtein, (1994). In 2nd ed. of book 'B Decays', S. Stone, ed., World Scientific.
56. M. Voloshin, TPI-MINN-00-17, UMN-TH-1851-00 (2000)
57. M. Neubert, C.T. Sachrajda, Nucl. Phys. **B483**, 339 (1997)
58. J.L. Rosner, Phys. Lett. B**379**, 267 (1996)
59. P. Abreu et al., Eur. Phys. J. **C10**, 185 (1999)
60. K. Ackerstaff et al., Phys. Lett. B**426**, 161 (1998)
61. R. Barate et al., Eur. Phys. J. **C2**, 197 (1998)
62. F. Abe et al., Phys. Rev. Lett. **77**, 1439 (1996)
63. V.M. Abazov et al., Phys. Rev. Lett. **99**, 182001 (2007)
64. V.M. Abazov et al., Phys. Rev. Lett. **94**, 102001 (2005)
65. V.M. Abazov et al., Phys. Rev. Lett. **99**, 142001 (2007)
66. A. Abulencia et al., Phys. Rev. Lett. **98**, 122001 (2007)
67. T. Aaltonen et al., Phys. Rev. Lett. **104**, 102002 (2010)

Chapter 6
Flavor Oscillations

Weakly decaying neutral mesons can oscillate into their anti-particles via second order weak interaction processes as shown for the B_s^0 meson in Fig. 6.1. The time evolution of such states is discussed in Chap. 3. In case of no CP violation and negligible decay width difference the probability density function for decays of non-oscillated meson states to a flavor-specific final state is obtained from Eqs. (3.43) and (3.46):

$$P_{n/osc}(t) = \Gamma e^{-\Gamma t}(1 + \cos \Delta m\, t). \tag{6.1}$$

For oscillated states Eqs. (3.44) and (3.45) lead to

$$P_{osc}(t) = \Gamma e^{-\Gamma t}(1 - \cos \Delta m\, t). \tag{6.2}$$

The oscillations cause a time-dependent asymmetry, $\mathscr{A}_{mix}(t)$, between decays of non-oscillated and oscillated meson states as defined in Eq. (3.47). In case of no detector effects the experimentally measured asymmetry would be

$$\mathscr{A}_{mix}(t) = \frac{N_{n/osc}(t) - N_{osc}(t)}{N_{n/osc}(t) + N_{osc}(t)} = \cos \Delta m\, t, \tag{6.3}$$

where $N_{n/osc}(t)$ ($N_{osc}(t)$) is the number of observed decays at a time t of a meson state with the same (opposite) flavor quantum number as it had at the time of production. The flavor quantum number of a heavy meson is given by the sign of the charge of the heavy quark, meaning that B mesons with b or $\bar{b}$ quark and charm mesons with c or $\bar{c}$ quark are distinguished.

Thus a measurement of the oscillation frequency Δm requires the experimental determination of the following properties for each reconstructed meson:

- the decay time, t,
- the flavor of the meson at production time, and
- the flavor of the meson at the time of its decay.

T. Kuhr, *Flavor Physics at the Tevatron*, Springer Tracts in Modern Physics 249,
DOI: 10.1007/978-3-642-10300-1_6,

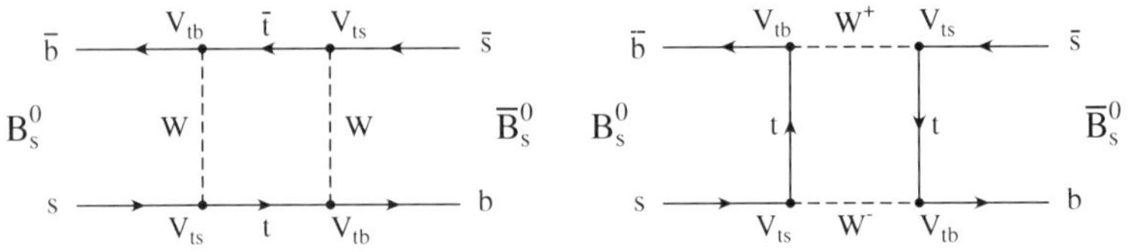

Fig. 6.1 Feynman diagrams of B_s^0 oscillations

The reconstruction of decay times from decay lengths and momenta is described in the previous Chapter. In flavor specific decays the meson flavor at decay time is unambiguously defined by the charges of the daughter particles. The task of determining the flavor at production time is called flavor tagging and described in the next Section.

6.1 Flavor Tagging

The flavor of a meson state at its production time has to be inferred from the particles in the event other than the decay products of the meson. Often this flavor assignment cannot be done unambiguously. To quantify the performance of flavor tagging algorithms, two variables are introduced. The efficiency

$$\varepsilon = \frac{N_{tag}}{N_{all}}, \tag{6.4}$$

where N_{tag} (N_{all}) is the number of tagged (all) events, specifies the fraction of events on which the tagging algorithm can be applied. The dilution measures how often the tagging decision is correct and is defined as

$$D = \frac{N_R - N_W}{N_R + N_W}, \tag{6.5}$$

where N_R (N_W) is the number of events with correct (wrong) tagging decision. The dilution is connected to the wrong tag fraction

$$w = \frac{N_W}{N_R + N_W} \tag{6.6}$$

by

$$D = 1 - 2w. \tag{6.7}$$

The name dilution is somewhat misleading because the higher the value of D, the more reliable is the tagging decision. The values of D range from -1 to 1, where $D = 1$ means perfect tagging, $D = 0$ indicates a tagger with random decision, and $D = -1$ tells that the decision is always opposite to the true flavor.

A consequence of the imperfect tagging is that only the fraction $1 - w$ of the non-oscillated events are reconstructed as non-oscillated. On the other hand, a fraction of w of the oscillated events are wrongly identified as non-oscillated. This leads to a probability density function for events with the same reconstructed production and decay flavor of

$$\begin{aligned} P_{n/osc}^{obs}(t) &= (1 - w) P_{n/osc}(t) + w P_{osc}(t) \\ &= \Gamma e^{-\Gamma t}(1 + \cos \Delta m\, t - w - w \cos \Delta m\, t + w - w \cos \Delta m\, t) \\ &= \Gamma e^{-\Gamma t}(1 + D \cos \Delta m\, t). \end{aligned} \tag{6.8}$$

In an analog way one gets

$$P_{osc}^{obs}(t) = \Gamma e^{-\Gamma t}(1 - D \cos \Delta m\, t) \tag{6.9}$$

leading to an observed asymmetry of

$$\mathscr{A}_{mix}^{obs}(t) = D \cos \Delta m\, t. \tag{6.10}$$

So the wrong tagging decisions cause a reduction of the amplitude of the observed oscillations by a factor D.

The dilution makes it more difficult to experimentally establish a non-zero amplitude of the $\cos \Delta m\, t$ term and thus a signal for flavor oscillations. In terms of statistical power, a data sample of N events with dilution D is equivalent to a smaller data sample of $D^2 N$ events with perfect tagging. A further reduction of the statistical power comes from the tagging efficiency ε defined in Eq. (6.4). Both factors are combined in the effective tagging efficiency εD^2. This is a good measure of the performance of a tagger, because the sensitivity to an oscillation signal of a data sample of N events with tagging efficiency ε and dilution D is the same as the sensitivity of a data sample of $\varepsilon D^2 N$ events with perfect tagging.

The values of the tagging performance variables, ε and D, do not only depend on the tagging algorithm, but also on the data sample. This fact can be exploited to increase the sensitivity of measurements by splitting the analyzed data sample into sub-samples with different tagger performance and using the sample-specific dilution instead of an average one. In the limit of infinite sub-samples this corresponds to the assignment of an individual dilution to each event. The principle of these event-by-event dilutions is similar to using event-by-event resolutions, for example in mass or decay time measurements. In the same way as mass or decay time uncertainties are measured for each event, the tagging dilution has to be determined for each event. The task of a flavor tagging algorithm is therefore not only to make a decision about the flavor, but also to estimate the dilution for a given event. This estimate is called predicted dilution, D_{pred}, and a calibration has to ensure that it reflects the true dilution.

Several properties of an event can be used to infer the flavor of a produced meson as illustrated for a B_s^0 meson in Fig. 6.2. The tagging algorithms employed at the

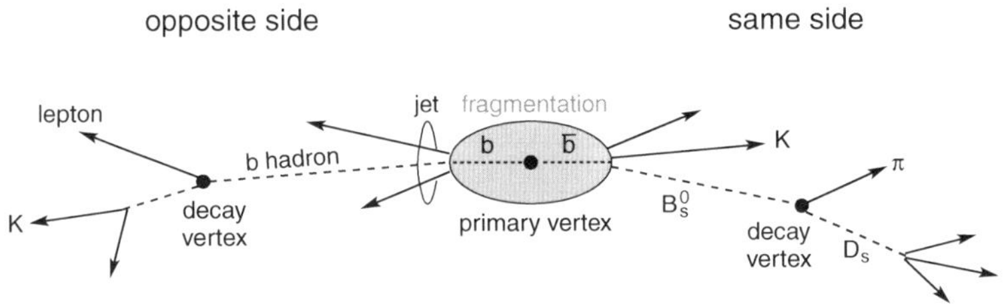

Fig. 6.2 Illustration of objects used for B_s^0 flavor tagging

Tevatron can be divided into two categories. The first category exploits the fact that b quarks are usually created in $b\bar{b}$ pair events. While one of the b quarks produces the (partially) reconstructed signal B meson, the other b quark fragments into a b hadron with opposite flavor of the signal B. Algorithms based on the identification of the flavor of the second b hadron in the event are called opposite-side taggers. They determine the flavor via the charge of the lepton from semileptonic b quark decays, the charge of kaons from $b \to c \to s$ quark transitions, or more inclusive quantities like the weighted charge of tracks forming a b hadron vertex or a b hadron jet.

Wrong tag decision can be caused by a wrong identification of the tagging objects, for example the lepton from $b \to c \to \ell$ decays has opposite charge to the lepton from the direct semileptonic decays. A further reduction of the performance of opposite-side taggers comes from B^0 and B_s^0 mesons that have oscillated. An advantage of opposite-side taggers is that they are independent of the type of the signal B meson because the two b quarks fragment independently. Therefore opposite-side taggers can be calibrated with B^+ and B^0 mesons and then applied to B_s^0 mesons.

The second category of taggers is based on the identification of particles produced in the fragmentation of the signal B meson and is called same-side taggers. To create a B_s^0 meson, an s quark is needed. The s quark is usually created in the fragmentation by a gluon that splits into an $s\bar{s}$ pair. If the $\bar{s}$ quark hadronizes into a charged kaon, the charge tags the flavor of the s quark in the kaon and by this also the flavors of the s and b quarks in the B_s^0 meson. So the B_s^0 same-side taggers determine the flavor by the charge of a kaon that is kinematically close to the B_s^0 signal meson.

Same-side taggers strongly depend on the type of meson that they should tag. In case of a B^0 meson the leading fragmentation partner is expected to be a pion instead of a kaon. Therefore it is not possible to calibrate a same-side tagger with one type of B mesons and then apply it to another type.

The categorization of taggers in opposite and same side suggests that two hemispheres, corresponding to a back-to-back topology of the two b quarks, can be identified in the experiment. While this was usually the case at LEP, a clear separation of the fragmentation and decay products of the two b quarks is often not possible at the Tevatron. The low momentum, compared to LEP, and production mechanisms like gluon splitting cause the two "hemispheres" to overlap. Moreover the particles produced by the second b quark may be outside the detector acceptance so that no opposite-side tag can be applied.

To avoid a contamination from same-side tracks in the opposite side tagging algorithms, CDF and D0 reject tracks around the B signal meson in the opposite side taggers. The rejection criteria is based on the angle or ΔR value between B meson and track, where $\Delta R = \sqrt{(\Delta\phi)^2 + (\Delta\eta)^2}$ and $\Delta\phi$ ($\Delta\eta$) is the difference in azimuthal angle (pseudorapidity). Jet clustering algorithms are applied to combine opposite side tracks that are assumed to stem from the fragmentation of the same parton.

The first kind of opposite side taggers used by CDF and D0 is based on muons. CDF identifies muons via a likelihood that is constructed from track, calorimeter, and muon detector information [1]. The predicted dilution is parametrized as a function of this likelihood and the momentum of the muon with respect to the jet that it contains,

$$p_T^{rel} = \frac{|\mathbf{p}(j) \times \mathbf{p}(\mu)|}{p(j)}, \tag{6.11}$$

where $\mathbf{p}(j)$ and $\mathbf{p}(\mu)$ are the momentum vectors of the jet and the muon, respectively. D0 selects the muon candidate with the highest number of hits in the muon detector and calculates the muon jet charge defined as

$$Q_{\mu j} = \frac{\sum_i q_i p_{T,i}}{\sum_i p_{T,i}}, \tag{6.12}$$

where the sum runs over all tracks within a cone of $\Delta R < 0.5$ around the muon, and q_i and $p_{T,i}$ are the charges and transverse momenta of the tracks, respectively [2]. The PDFs of $Q_{\mu j}$ for both flavors, $f_{\mu j}^{b/\bar{b}}(Q_{\mu j})$, are determined from data in a calibration procedure described below. The predicted dilution is obtained from a likelihood ratio of both PDFs:

$$D_{pred,\mu j} = \frac{1 - r_{\mu j}}{1 + r_{\mu j}}, \tag{6.13}$$

with

$$r_{\mu j} = \frac{f_{\mu j}^{\bar{b}}(Q_{\mu j})}{f_{\mu j}^{b}(Q_{\mu j})}. \tag{6.14}$$

In a similar way, CDF and D0 construct electron based taggers [2, 3]. Both experiments use track, calorimeter, and preshower detector information to identify electrons. In addition CDF uses specific energy loss in the drift chamber and the Central Electromagnetic Shower detector. The predicted dilution of the CDF electron tagger is determined as a function of the electron likelihood, $p_{T,rel}$, and a flag that indicates whether the track circle in the transverse plane contains the primary vertex or not. The latter variable helps to identify electron from conversion. D0 calculates the predicted dilution of the electron tagger from an electron jet charge, Q_{ej}, defined in the same way as the muon jet charge in Eq. (6.12).

CDF has also developed an opposite-side kaon tagger. A kaon track is identified using the dE/dx value measured in the drift chamber and the time of flight measurement. The dilution is predicted as a function of kaon momentum relative to the jet momentum for events with a secondary vertex, or from the kaon track impact parameter significance otherwise.

The last kind of opposite-side taggers employed by CDF is the jet charge tagger [4]. In a first step it uses a neural network to determine the probability of a track to originate from a b hadron decay, regardless of its flavor. The input to the network includes information like the distance from the primary vertex and the momentum with respect to the jet. In the second step, jet properties are calculated from the track probabilities and combined in further neural networks for three different classes of jets. The classes are defined via the presence of a secondary vertex candidate and a track with high probability to originate from a b hadron decay. The jet with the highest network output is selected as tagging jet. For this jet, the jet charge,

$$Q_j = \frac{\sum_i q_i p_{T,i}(1 + t_i)}{\sum_i p_{T,i}(1 + t_i)}, \tag{6.15}$$

is calculated, where the sum runs over all tracks in the jet and t_i is the track probability obtained in the first step. The predicted dilution is parametrized for each of the three jet classes by a linear function of the jet charge.

Instead of a jet charge tagger, D0 employs a secondary vertex charge tagger [2]. In case of a vertex that has at least two tracks and is significantly displaced from the primary vertex, the secondary vertex charge is calculated in the following way:

$$Q_{SV} = \frac{\sum_i (q_i p_{L,i})^k}{\sum_i p_{L,i}^k}, \tag{6.16}$$

where the sum runs over all tracks coming from the secondary vertex and $p_{L,i}$ is the momentum of the i-th track in direction of the combined vertex momentum. The parameter k is set to 0.6 to achieve optimal tagger performance. Like in the case of the lepton tagger, the prediction dilution is obtained from a likelihood ratio in the charge variable.

Finally, D0 also considers the event charge,

$$Q_{EV} = \frac{\sum_i q_i p_{T,i}}{\sum_i p_{T,i}}, \tag{6.17}$$

where the sum is taken over all opposite-side tracks.

The order in which the taggers were described above, is roughly in increasing order of efficiency. While the fraction of events with exclusively reconstructed final state objects, like muons or electrons, is low, an opposite-side jet can be found in almost all events. On the other hand the tagging decision of the lepton taggers is more reliable than the decision of more inclusive taggers, like the event charge tagger. The effective tagging efficiency of the individual taggers is of the order of 1 %.

The taggers are not mutually exclusive. If more than one tagger provides a decision of a given event, two approaches are possible to deal with this situation. The first solution is to pick one of the taggers.

Such a hierarchical combination of taggers was initially used by CDF. If an event was tagged by the muon tagger, its decision was taken because the muon tagger has the highest average dilution of all taggers. In case of no muon tag the electron tag was taken and if no lepton tag was available, the jet charge tagger was the final choice.

D0 combines the decision of two taggers by multiplying their likelihood ratios, but still keeps some hierarchy. The combinations of taggers with decreasing priority are muon and secondary vertex charge, electron and secondary vertex charge, and secondary vertex and event charge. If there are correlations between the taggers the predicted dilution calculated from the likelihood ratio product according to Eq. (6.13) does not correspond any more to the true dilution. An alternative combination was developed using multidimensional likelihood functions. In addition to the charges, variables like (relative) momenta and impact parameter significances were considered. However, the issue of the construction of a likelihood function is that the number of events per bin in the multidimensional space rapidly decreases with increasing dimensionality. Therefore the dimension was limited to three and no improvement with respect to the default combination was achieved.

While the multidimensional likelihood approach gives theoretically the best result, the practical issue of limited statistics is solved better by other multivariate techniques. CDF uses a neural network [5] to combine the decisions of the individual opposite-side taggers. The network can deal with undefined inputs, which is important for events where not all tagger decisions are available. The correlations between the taggers, which are ignored by the hierarchical combination, are automatically taken into account by the network and lead to a relative increase of the effective tagging power of about 20 %. A further nice feature of an optimally trained neural network is that its output value, $o_i \in [0; 1]$, is equal to the probability of the event i to be signal as defined in the training. As the tagger combination network is trained to separate the two flavors, its output is directly the probability of an event to have a particular flavor. The tagging decision is thus given by the sign of $(o_i - 0.5)$ and the predicted dilution is $D_{pred} = 2|o_i - 0.5|$.

CDF and D0 calibrate their opposite-side taggers on measured data. D0 selects $B \to D^0\mu^+X$ and $B \to D^{*-}\mu^+X$ events and applies the taggers on the opposite side. The former sample is dominated by B^+ decays, the latter by B^0 decays. The exact sample composition is determined in detail from simulation. This allows to calculate the visible proper decay time distribution of non-oscillated and oscillated events and their asymmetry. The flavor at decay time is given by the charge of the muon. The flavor at production time is experimentally determined by the opposite-side tagging decision. Both tags combined yield the observed asymmetry as a function of t_{vis} which is fitted by a function that includes the dilution factor.

This calibration of the dilution is performed for sub-samples in bins of predicted dilution. The measured dilution is then parametrized as a function of predicted dilution to obtain an event-by-event calibrated dilution. The result of this calibration on a data sample of 8 fb^{-1} is shown in the Fig. 6.3. The effective tagging

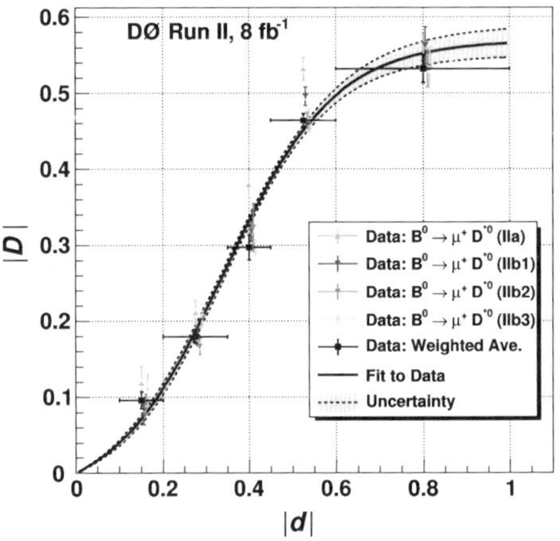

Fig. 6.3 Opposite site tagger calibration [2]. The plot shows the measured dilution $|D|$ versus the predicted dilution $|d|$ for different data taking periods and the average together with the fitted parametrization

efficiency of the combined tagger on the calibration sample is $\varepsilon D^2 = (2.48 \pm 0.21\,(\text{stat})^{+0.08}_{-0.06}\,(\text{syst}))\,\%$ [2].

CDF uses $B^+ \to J/\psi K^+$ events for the opposite-side tagger calibration. Because B^+ mesons do not oscillate, the kaon charge directly determines the production flavor. So the number of correctly and wrongly tagged events is easily obtained by the comparison of kaon charge and tagger decision. As the neural network for the combined tagger was trained on real data, it should already be well calibration. A remaining sample dependence of the calibration is taken into account by a scale factor that is multiplied to the predicted dilution. The scale factor is determined in a fit to the measured dilution as a function of predicted dilution. Since the tagger could in principle perform differently for B^+ and B^- mesons, the scale factor is determined separately for both samples. But the two measured scale factors are both consistent with 1 indicating that the tagger was already calibrated well on the data sample used for the network training. The plots in Fig. 6.4 show the calibration for both flavors on a data set of 5.2 fb^{-1}. The effective tagging efficiency of the combined opposite-side tagger is quoted as $\varepsilon D^2 = (1.8 \pm 0.1)\,\%$ from the calibration for the B_s^0 oscillation measurement [6]. The lower value compared to D0 is mainly caused by the smaller muon detection acceptance of the CDF detector.

In addition to the combined opposite-side tagger, CDF also employs a same-side tagging algorithm. All tracks in a cone of $\Delta R < 0.7$ around the momentum direction of the signal B meson are considered. The version of the tagger that was used for the first B_s^0 mixing measurement selected the track with the highest likelihood to be a kaon as tagging track [8]. The kaon likelihood is determined from the specific energy loss in the drift chamber and the time of flight measurement. The flavor tagging decision is given by the charge of the kaon track. The predicted dilution is parametrized as a function of the kaon likelihood variable for two classes of events where either all tracks in the $\Delta R < 0.7$ cone have the same charge or not. Because the tagging algorithm is based on the identification of the kaon from the fragmentation it is called same-side kaon tagger (SSKT).

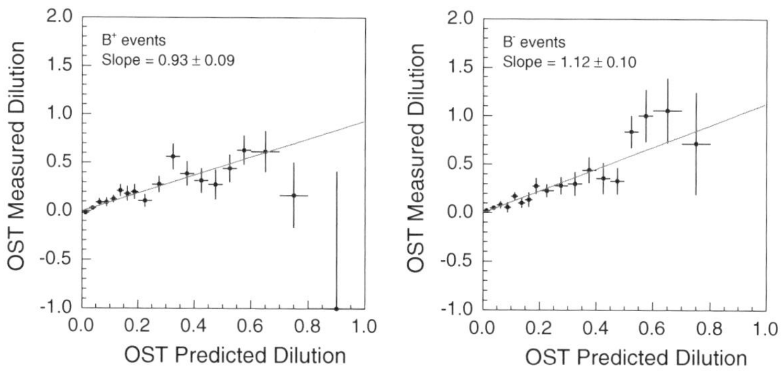

Fig. 6.4 Opposite side tagger calibration on a B^+ (*left*) and B^- data sample (*right*) [7]

To improve the performance of the SSKT, the kaon likelihood variable is combined with further kinematic variables in a neural network [6]. These variables include the transverse momentum of the track, the momentum perpendicular and in direction of the B meson, and ΔR with respect to the B meson. The network output is used to select the tagging track and to parametrize the predicted dilution.

For the B_s^0 oscillation measurements described in the next Section, the SSKT predicted dilution was calibrated with simulated data. To verify that the simulation describes the data and in particular the fragmentation process well, the tagger performance is measured on B^+ and B^0 data and compared to the results on simulated data. Although the tagger depends on the type of B meson and is not expected to yield a reliable dilution prediction if applied to another B meson type, it is assumed that a comparison of tagger results on data and simulation is sensitive to a mismodeling in the simulation. The effective tagging efficiency for the sample of hadronically (semileptonically) decaying B_s^0 mesons selected in the B_s^0 oscillation analysis is determined to be 3.7 % (4.8 %) [6]. Compared to the same-side kaon tagger using only particle identification information this is a relative improvement of about 10 %. The relative uncertainty on εD^2, as estimated from the comparison of results on real and simulated data, is approximately 25 %. The different values for the two samples are a result of the momentum dependency of the tagger performance.

Unlike the opposite-side taggers, the SSKT cannot be calibrated on B^+ or B^0 data and then applied to B_s^0 data. The only way to calibrate the SSKT on data is to perform a complete B_s^0 oscillation analysis as described in the next Section. Such an analysis was performed on a data sample of 5.2 fb^{-1} [7]. The obtained correction factor for the predicted dilution is 0.94 ± 0.15 (stat) $\pm\, 0.13$ (syst), well consistent with 1. The measured B_s^0 mixing frequency of $\Delta m_s = (17.79 \pm 0.07\ \text{(stat)})\ \text{ps}^{-1}$ is in good agreement with the previous CDF measurement [6] and consistent with the standard model expectation [9, 10]. The amplitude plot (explained below) determined in the calibration of the SSKT on data is shown in Fig. 6.5.

The performance of the SSKT is higher than that of the combined opposite-side tagger because the kaon from the fragmentation has a higher probability to

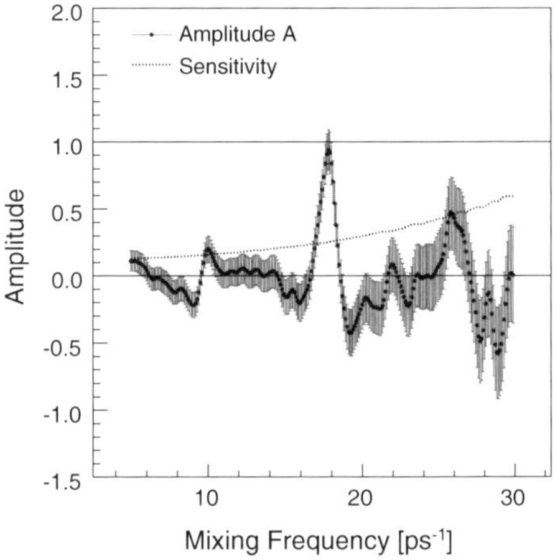

Fig. 6.5 Amplitude scan of the B_s^0 oscillation measurement for the same-side kaon tagger calibration [7]. The amplitude at the B_s^0 oscillation frequency of $\Delta m_s = 17.79\ \text{ps}^{-1}$ yields the calibration factor $0.94 \pm 0.15\ (\text{stat}) \pm 0.13\ (\text{syst})$

be within the detector acceptance than the opposite-side objects since the kaon is kinematically close to the already reconstructed B_s^0 signal meson. If both tagger decisions are available in an event, they are combined under the assumption that they are independent:

$$\xi D = \frac{(1 + \xi_{\text{OS}} D_{\text{OS}})(1 + \xi_{\text{SS}} D_{\text{SS}})}{2} - 1, \tag{6.18}$$

where ξ is the flavor of -1 or $+1$ and D the predicted dilution of the combined tagger decision. The corresponding variables for the opposide- and same-side tagger are denoted by the subscripts OS and SS, respectively. The assumption of independence of same-side and opposite-side taggers is reasonable, because they use disjunct collections of tracks. The independence was also verified with simulation.

6.2 B_s^0 Oscillations

To establish a B_s^0 oscillation signal and to measure Δm_s, the time-dependent decay rate of B_s^0 mesons to a flavor-specific final state is fitted with a PDF derived from Eqs. (6.1) and (6.2). If no significant signal is observed, a lower limit on the mixing frequency is determined. A method that is well suited to combine limits of different experiments and to visualize the sensitivity of the measurements is the amplitude scan [11]. The basic idea is to perform a Fourier transformation. Technically this is done by introducing an amplitude, A, in Eqs. (6.1) and (6.2) and fitting $A(\Delta m)$ by minimizing the constructed likelihood function for a fixed value of Δm:

$$P^{obs}_{n/osc}(t) = \Gamma e^{-\Gamma t}(1 + AD\cos\Delta m\, t), \tag{6.19}$$

$$P^{obs}_{osc}(t) = \Gamma e^{-\Gamma t}(1 - AD\cos\Delta m\, t). \tag{6.20}$$

Since the B_s^0 mesons oscillate with a fixed frequency, Δm_s, the amplitude of the Fourier transformed distribution peaks at this value. In the amplitude scan the height of the peak at the true frequency is normalized such that the expectation value is 1 when all detector effects are properly taken into account. For frequencies well apart from the true value of Δm_s, the amplitude is expected to be consistent with zero.

Frequencies at which the amplitude is inconsistent with 1 can be excluded. For a 95 % confidence level limit this corresponds to

$$A + 1.645\,\sigma_A < 1, \tag{6.21}$$

where σ_A is the uncertainty of the amplitude. The sensitivity of an experiment, $\Delta m_s^{\rm sens}$, is defined by the Δm_s value up to which it is expected to exclude an amplitude of 1 and thus B_s^0 oscillations in case of no signal:

$$1.645\,\sigma_A = 1. \tag{6.22}$$

An oscillation signal is observed if $A = 0$ can be excluded with sufficient confidence

$$\frac{A}{\sigma_A} > n, \tag{6.23}$$

where n is the desired number of standard deviations.

Several factors limit the experimental sensitivity. One of them is the finite number of signal events, S. This effect alone leads to an expected significance of the oscillation signal of $(A/\sigma_A)_{\rm exp} = \sqrt{S/2}$, corresponding to the significance with which the amplitude of a cos oscillation can be measured with S events. The contamination of the data sample with background events reduces the expected significance further by a factor of $\sqrt{S/(S+B)}$, where B is the background yield. As already discussed in the previous Section, an imperfect tagging effectively scales the signal yield down by a factor εD^2. Finally, the decay time resolution dilutes the oscillation signal. Assuming a Gaussian resolution function with width σ_t, the Fourier transformation in the amplitude scan converts this into an exponential function. With the factor $\exp[-(\Delta m_s \sigma_t)^2/2]$ from the decay time resolution the overall expected signal significance is to a good approximation [11]

$$\left(\frac{A}{\sigma_A}\right)_{\rm exp} = \frac{S}{\sqrt{S+B}}\sqrt{\frac{\varepsilon D^2}{2}}\, e^{-(\Delta m_s \sigma_t)^2/2}. \tag{6.24}$$

The formula shows that the dependency of the experimental sensitivity on Δm_s is driven by the decay time resolution. This explains the different sensitivity dependency of partially reconstructed semileptonic decays and fully reconstructed hadronic

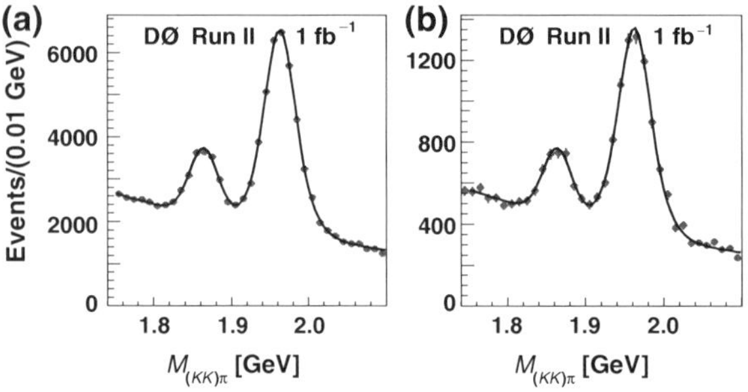

Fig. 6.6 Invariant mass distribution of D_s^- candidates for all selected events (*left*) and flavor-tagged events (*right*) [16]

decays. At low Δm_s values where $\Delta m_s \sigma_t$ is small and therefore the resolution factor, $\exp[-(\Delta m_s \sigma_t)^2/2]$, close to 1, the semileptonic decays provide a better sensitivity because of the higher signal yield. For increasing Δm_s values the resolution factor decreases faster for semileptonic decays so that at higher Δm_s values the hadronic decays have a superior sensitivity.

Attempts to resolve the B_s^0 oscillations have been made at LEP [12–14] and SLD [15], but resulted only in lower limits on Δm_s. The first indication of a finite B_s^0 oscillation frequency was reported by D0 using a data sample of 1 fb^{-1} [16]. B_s^0 mesons are partially reconstructed in the decay $B_s^0 \to D_s^- \mu^+ X$ with $D_s^- \to \phi\pi^-$ and $\phi \to K^+K^-$. Kinematic and vertex fit quality variables are combined in a likelihood ratio and the requirement on this ratio is chosen to maximize the expected signal significance, $S/\sqrt{S+B}$, where S and B are the predicted signal and background yields, respectively. The opposite-side tagger described in the previous Section is applied to the selected sample. Figure 6.6 shows the invariant mass distribution of the selected D_s^- candidates. The left peak is caused by $D^- \to K^+\pi^-\pi^-$ decays where one of the pions is misidentified as a kaon.

The data is fitted with a likelihood function that depends on the following observables: the visible proper decay length, its uncertainty, the D_s^- candidate invariant mass, the predicted dilution, and the likelihood ratio selection variable. Contributions from $D_s^-\mu^+$, $D^-\mu^+$, and combinatorial background are considered. The PDFs of the observables other than the visible proper decay length are determined from data and assumed to be independent. The composition of processes contributing to the $D_s^-\mu^+$ and $D^-\mu^+$ signals is determined from simulation. It includes B meson decays to two charm mesons with a sequential semileptonic D meson decay and $c\bar{c}$ and $b\bar{b}$ events where D meson and muon originate from different initial quarks. The visible proper decay length PDF is constructed from the theoretical distribution of the known sample contributions convolved with a resolution function. The resolution is parametrized by a double Gaussian function with widths given by the measured uncertainty scaled by two factors. The scale factors are determined on a sample of $J/\psi \to \mu^+\mu^-$ decays. The visible proper decay length PDF of the combinatorial background takes

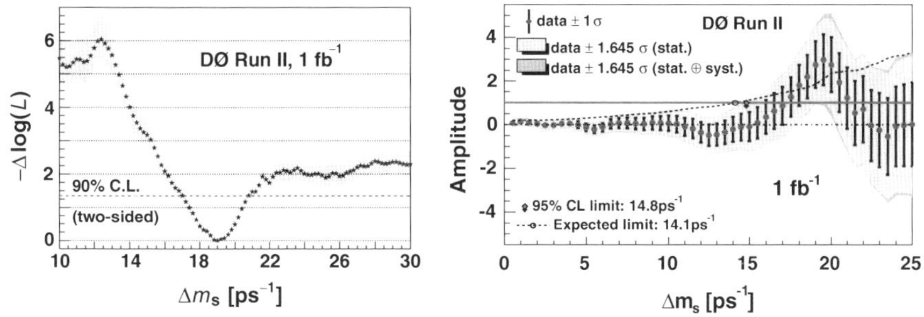

Fig. 6.7 Likelihood profile of the B_s^0 oscillation fit (*left*) and amplitude scan (*right*) [16]. The shaded band in the left plot represents the systematic uncertainty

contributions from prompt events and long-lived events into account, where the latter component includes a fraction of oscillating events with frequency Δm_d.

The result of the fit is shown in Fig. 6.7. The likelihood function has a minimum at 19 ps^{-1} and a 90 % confidence level interval of 17 ps^{-1} $< \Delta m_s <$ 21 ps^{-1} is derived [16]. The significance of the signal is assessed by determining the probability to obtain a minimum as deep or deeper than the observed one in the range 16 ps^{-1} $<$ $\Delta m_s <$ 22 ps^{-1} if the true frequency is higher than 22 ps^{-1}. The range is chosen to cover the frequencies between the limit from previous experiments and the value where the experimental sensitivity ends and the likelihood profile reaches a plateau. A probability of (5.0 ± 0.3) % is estimated with an ensemble test where the fit was repeated on data several times with randomized tagger decision. The randomization of the the tagger decision leads to a dilution of zero so that the expected amplitude is zero as well.

The indication of the B_s^0 oscillation signal is also seen in the amplitude plot. At $\Delta m_s = 19$ ps^{-1} the measured amplitude is 2.5 standard deviations away from zero. It is consistent with the oscillation hypothesis of $A = 1$ at the 1.6σ level. The fact that the amplitude value is much larger, but consistent with one, suggests that a statistical fluctuation to a higher A value happened in this experiment. Therefore it was possible to see an indication of a signal at the $\sim 2\sigma$ level at a frequency which is above the sensitivity, as defined in Eq. (6.22), of $\Delta m_s = 14.1$ ps^{-1}.

Briefly after the D0 result was reported, CDF presented the first evidence of a B_s^0 oscillation signal and the first measurement of Δm_s using a data sample of 1 fb^{-1} [8]. In addition to semileptonic decays $B_s^0 \to D_s^- \ell^+ X$, where $\ell = \mu$ or e, also hadronic B_s^0 decays to $D_s^- \pi^+$ and $D_s^- \pi^+ \pi^+ \pi^-$ are reconstructed. The D_s^- meson is reconstructed in the decay modes $\phi(\to K^+K^-)\pi^-$, $K^{*0}(\to K^+\pi^-)K^-$, and $\pi^-\pi^+\pi^-$. To avoid a contamination in the $K^{*0}K^-$ mode of $D^- \to K^+\pi^-\pi^-$ decays with a pion misidentified as a kaon, events that are consistent with the D^- decay hypothesis are rejected. The hierarchical combination of opposite-side taggers and the same-side tagger based only on particle identification are applied.

The oscillation frequency is determined in a maximum likelihood fit of B_s^0 candidate invariant mass, decay time, its uncertainty, and tagger decision and pre-

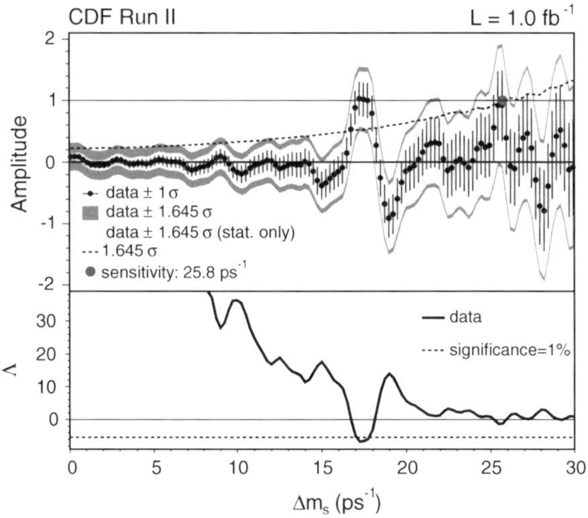

Fig. 6.8 Amplitude scan and likelihood ratio of the B_s^0 oscillation fit [8]

dicted dilution. In case of semileptonic decays the invariant mass of the $D_s^- \ell^+$ pair, $m_{D_s \ell}$, and the visible proper decay time are used. To improve the visible proper decay time resolution, the K factor distribution is determined in bins of $m_{D_s \ell}$ as shown in Fig. 6.10 for the updated CDF measurment. For reconstructed masses close to the true B_s^0 mass the missing momentum carried by the neutrino is small so that the K factor distribution is narrow and peaks close to 1. For the hadronic decays, the decay time dependent trigger selection efficiency has to be considered in the likelihood function. This is done by an efficiency curve derived from simulation. The decay time resolution is modeled by a Gaussian function with a width that is given by the measured uncertainty of each candidate. To account for possible differences between the measured and true uncertainty, the width is scaled by a factor that is calibrated as a function of kinematic and topological variables on a sample of promptly produced D mesons.

The PDFs of combinatorial background are determined from mass sidebands. Templates for B meson decays other than the $B_s^0 \to D_s^{-(*)} \ell^+ \nu$ contribution to the semileptonic sample are obtained from simulation. A sample of D_s^- combined with a track is used to estimate the contribution from events where a hadron is misidentified as a lepton. The mass and decay time models as well as the tagging variable distributions of the background are determined in separate fits and fixed in the final fit of the oscillation frequency.

The result of the fit is shown in Fig. 6.8 in form of an amplitude scan and a log likelihood ratio, Λ, between the hypothesis of an oscillation signal ($A = 1$) and no signal ($A = 0$):

$$\Lambda = \log\left(\frac{\mathscr{L}(A=1)}{\mathscr{L}(A=0)}\right). \tag{6.25}$$

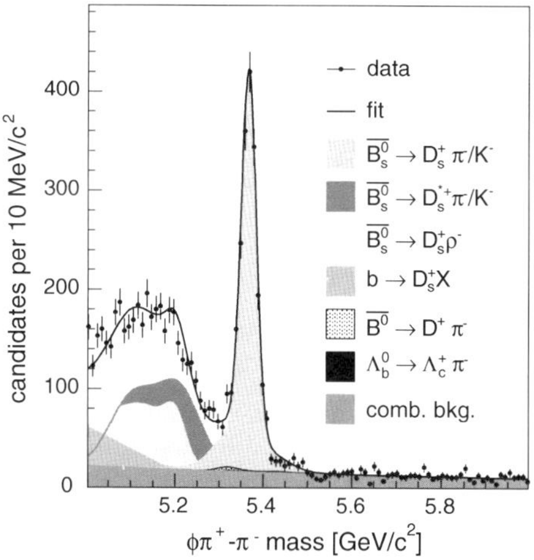

Fig. 6.9 Invariant mass distribution of $B_s^0 \to D_s^- \pi^+$ candidates with $D_s^- \to \phi\pi^-$ [6]

At $\Delta m_s = 17.3\ \text{ps}^{-1}$ the signal hypothesis is strongly favored with $\Lambda = -6.75$ and the minimization of the likelihood function with A fixed to 1 yields a measurement of $\Delta m_s = (17.31^{+0.33}_{-0.18}\ (\text{stat}) \pm 0.07\ (\text{syst}))\ \text{ps}^{-1}$ [8]. The systematic uncertainty is dominated by the uncertainty of the decay time measurement due to possible biases in the vertex reconstruction and the silicon detector alignment. The amplitude at $\Delta m_s = 17.3\ \text{ps}^{-1}$ is $A = 1.03 \pm 0.28$ (stat), 3.7 standard deviations away from zero and well consistent with 1 as expected for a B_s^0 oscillation signal and properly calibrated decay time resolution and flavor tagging algorithms. The probability that the observed signal is caused by a statistical fluctuation is determined by repeating the fit many times with random tagging decisions and looking for a value of $\Lambda < -6.75$ at any value of Δm_s. The low probability of 0.2 % justifies to interpret this as a B_s^0 oscillation signal, but is not yet sufficient to claim an observation.

To establish a clear B_s^0 oscillation signal, CDF analyzed the same dataset again with improved analysis techniques [6]. The first improvement is a more powerful B_s^0 signal selection. The use of particle identification information allows to drop the veto against $D^- \to K^+\pi^-\pi^-$ decays in the $D_s^- \to K^{*0}K^-$ mode. The selection is further optimized by combining the selection variables in a neural network. These improvements lead to an increase in the signal yield from 3,600 to 5,600 fully reconstructed hadronic decays and from 37,000 to 61,500 partially reconstructed semileptonic decays compared to the previous analysis.

As can be seen in Fig. 6.9, partially reconstructed B_s^0 decays populate the invariant mass region below the nominal B_s^0 mass. These decays were excluded in the previous analysis. In the updated analysis the decays $B_s^0 \to D_s^{*-}\pi^+$ with $D_s^{*-} \to D_s^-\gamma/\pi^0$ and $B_s^0 \to D_s^-\rho^+$ with $\rho^+ \to \pi^+\pi^0$ are exploited to increase the signal yield in the mode $D_s^- \to \phi\pi^-$ by 3,100 events. Because the momentum of the not reconstructed photon or π^0 in the B_s^0 rest frame is small, the average decay time resolution

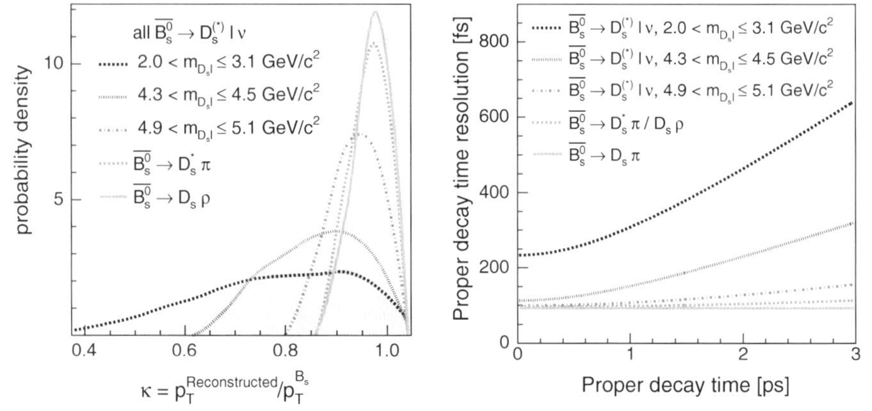

Fig. 6.10 K-factor distribution (*left*) and decay time resolution (*right*) for semileptonic decays in bins of invariant $D_s^- \ell^+$ mass and for (partially) reconstructed hadronic decays [6]

of 97 fs is only slightly worse than the resolution of fully reconstructed hadronic decays of 87 fs. As shown in Fig. 6.10 the resolution is also nearly independent of the decay time.

The last improvement compared to the previous analysis is achieved by the use of neural networks to combine the opposite-side tags and to improve the same-side kaon selection as discussed in Sect. 6.1.

The result of the fit is illustrated in Fig. 6.11. The amplitudes for semileptonic and hadronic modes are both consistent with a B_s^0 oscillation signal ($A = 1$) at $\Delta m_s \approx 17.75\ \text{ps}^{-1}$. The combined result yields an amplitude of $A = 1.21 \pm 0.20$ at $\Delta m_s \approx 17.75\ \text{ps}^{-1}$, again consistent with unity. The probability of a background fluctuation is determined to be 8×10^{-8}, corresponding to a significance of 5.4σ. A further measure of the improvement due to the more sophisticated analysis techniques is the increase of the Δm_s sensitivity from 25.8 to 31.3 ps^{-1}. The high sensitivity comes from the hadronic modes which alone provide a sensitivity of 30.7 ps^{-1}. The sensitivity of the semileptonic decays is 19.3 ps^{-1}.

The measured oscillation frequency is $\Delta m_s = 17.77 \pm 0.10\ (\text{stat}) \pm 0.07\ (\text{syst})$ ps^{-1} [6]. To illustrate that the data really shows a time dependent asymmetry between non-oscillated and oscillated B_s^0 states as defined in Eq. (6.3), the likelihood fit is performed in bins of decay time modulo the measured oscillation period. The fitted asymmetry is presented in Fig. 6.12. It nicely follows the expected cosine shape.

As can be seen from the Feynman diagram in Fig. 6.1, the B_s^0 oscillation frequency measurement is sensitive to the CKM matrix element V_{ts}. A detailed theoretical calculation yields [17]

$$\Delta m_s = \frac{G_F^2}{6\pi^2} \eta_B m_{B_s^0} \hat{B}_{B_s^0} f_{B_s^0}^2 m_W^2 S\left(\frac{m_t^2}{m_W^2}\right) |V_{tb} V_{ts}^*|^2, \tag{6.26}$$

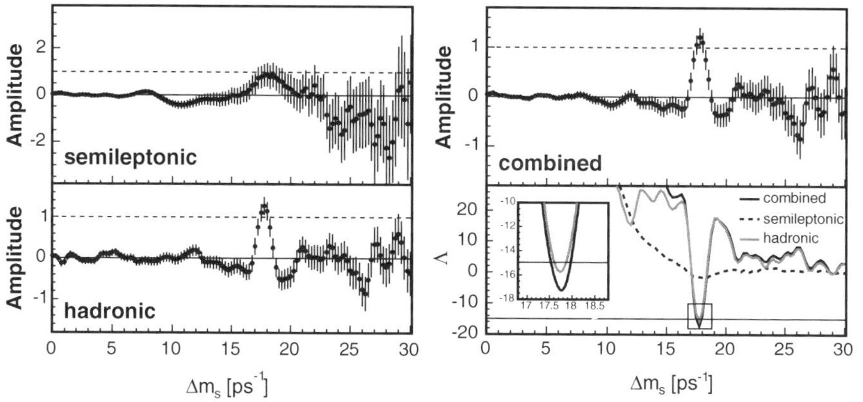

Fig. 6.11 Amplitude scans separately for semileptonic and hadronic modes (*left*) and combined together with the log likelihood ratio (*right*) [6]

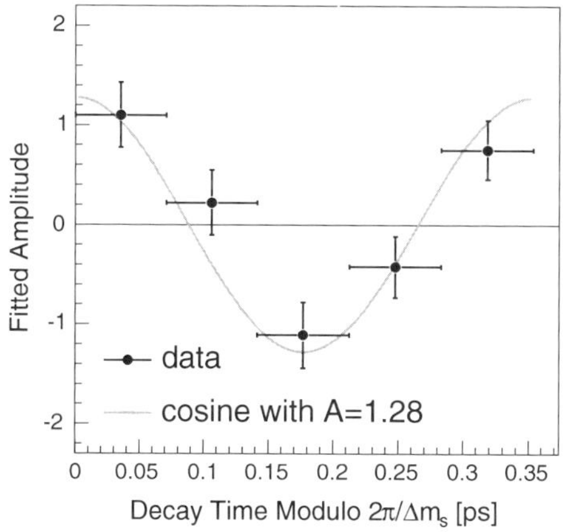

Fig. 6.12 Fitted asymmetry between non-oscillated and oscillated B_s^0 states [6]

where G_F is the Fermi constant, $m_{B_s^0}$, m_W, and m_t are the masses of the B_s^0 meson, the W boson, and the top quark, respectively, η_B is a numerical factor containing perturbative QCD corrections [18], and S is the Inami-Lim function [19]. Long distance QCD effects are parametrized by the B_s^0 meson decay constant, $f_{B_s^0}$, and the bag parameter, $\hat{B}_{B_s^0}$, which corrects for the neglection of interactions between the B_s^0 and $\bar{B}_s^0$ states in the calculation.

While it is in principle possible to extract $|V_{tb}V_{ts}^*|$ from the Δm_s measurement with Eq. (6.26), the precision would be rather limited by the theoretical uncertainties of the involved factors. Therefore it is advisable to consider the ratio of oscillation frequencies between B_s^0 and B^0 mesons, where several factors cancel:

$$\frac{\Delta m_s}{\Delta m_d} = \frac{m_{B_s^0}}{m_{B^0}} \frac{\hat{B}_{B_s^0} f_{B_s^0}^2}{\hat{B}_{B^0} f_{B^0}^2} \left| \frac{V_{ts}}{V_{td}} \right|^2 = \frac{m_{B_s^0}}{m_{B^0}} \xi^2 \left| \frac{V_{ts}}{V_{td}} \right|^2 \tag{6.27}$$

with

$$\xi = \frac{f_{B_s^0} \sqrt{\hat{B}_{B_s^0}}}{f_{B^0} \sqrt{\hat{B}_{B^0}}}. \tag{6.28}$$

With the value of ξ taken from Ref. [20], $\Delta m_d = 0.507 \pm 0.004\,\text{ps}^{-1}$ and $m_{B_s^0}/m_{B^0} = 1.01644$ from Ref. [21], the following ratio of absolute CKM matrix elements is obtained [6]:

$$\begin{aligned} \left| \frac{V_{td}}{V_{ts}} \right| &= \xi \sqrt{\frac{\Delta m_d m_{B_s^0}}{\Delta m_s m_{B^0}}} \qquad (6.29) \\ &= 0.2061 \pm 0.0007 (\Delta m_s) \pm 0.0008\ (\Delta m_d)^{+0.0080}_{-0.0060}\ (\xi). \end{aligned}$$

Although the theoretical uncertainties on the ξ parameter are only 3–4 %, they are still about an order of magnitude larger than the contributions from the experimental uncertainties on the B_s^0 and B^0 oscillation frequencies. It is also worth noting that the first direct observation of B_s^0 oscillations yielded a Δm_s measurement with a relative uncertainty that is slightly smaller than the uncertainty of the world average Δm_d value. Recently the LHCb experiment has presented a Δm_s measurement [22] with similar precision as the CDF result.

6.3 D^0 Oscillations

About one year after the observation of B_s^0 oscillations, also evidence for mixing of the D^0 meson was reported by Belle [23], BaBar [24] and CDF [25]. This confirmed the theoretical expectation that all weakly decaying neutral mesons state oscillate into their anti-particle.

Because of the GIM mechanism [26], the short-range contributions to D^0 mixing from box diagrams like the one in Fig. 6.1 for B_s^0 mesons are strongly suppressed in the standard model. The dominant contribution comes from long-range processes where the D^0 evolves into a $\bar{D}^0$ meson via virtual intermediate states like $\pi^+\pi^-$ as shown in Fig. 6.13. Such processes are hard to calculate which makes the interpretation of a D^0 oscillation signal in terms of long-range and new physics contributions difficult. In the standard model a slow oscillation and a small decay width difference with the parameters $x = \Delta m/\Gamma$ and $y = \Delta\Gamma/2\Gamma$ at the level of 1 % or below is predicted [27–31].

Like in the case of B_s^0 oscillations, D^0 oscillations can be observed experimentally by a time-dependent measurement of the decay rates of oscillated and non-oscillated states. The decay mode considered by CDF in an analysis of 1.5 fb^{-1} of data is

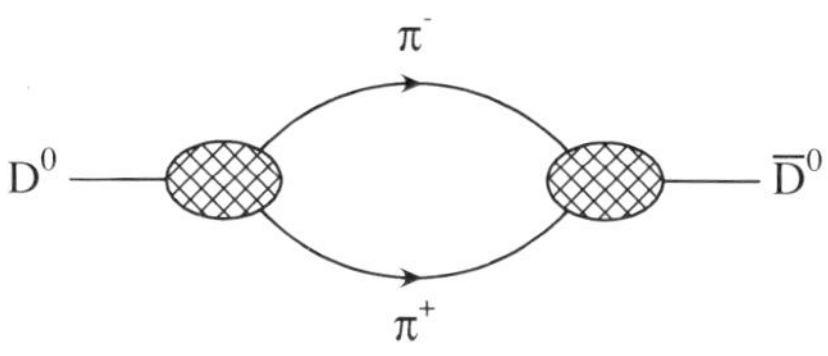

Fig. 6.13 Long distance diagram of a D^0 oscillation

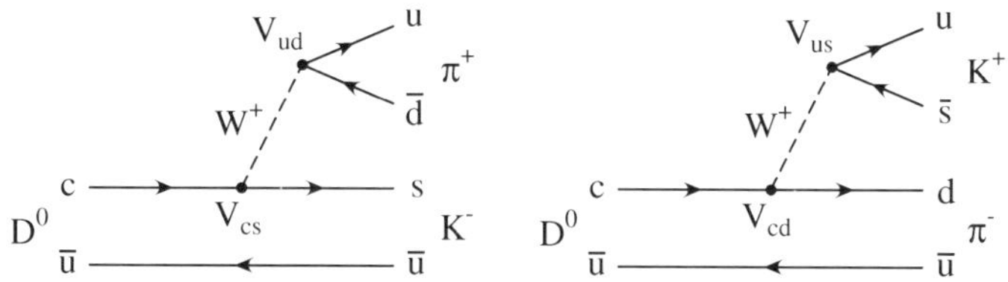

Fig. 6.14 Feynman diagrams of the Cabibbo-favored decay $D^0 \to K^-\pi^+$ (*left*) and the doubly Cabibbo-suppressed decay $D^0 \to K^+\pi^-$ (*right*)

$D^0 \to K^-\pi^+$ [25]. The basic idea of the analysis is to use the kaon and pion charges of this Cabibbo-favored decay, denoted as "right sign" (RS), as tag of the decay flavor. But the decay is not a pure flavor specific mode. As illustrated in Fig. 6.14, the "wrong sign" (WS) charge combination $K^+\pi^-$ is not only accessible by D^0 mesons via an oscillation to a $\bar{D}^0$ meson, but also via a doubly Cabibbo-suppressed decay. The ratio of decay amplitudes is characterized by the parameter R_D,

$$R_D = \left| \frac{A_{\bar{f}}}{A_f} \right|^2, \tag{6.30}$$

where f and $\bar{f}$ denote the Cabibbo-favored and doubly Cabibbo-suppressed final state, respectively. From the CKM matrix elements involved in the tree level decays one can obtain the following expectation for R_D:

$$R_D = \left| \frac{V_{cd} V_{us}^*}{V_{cs} V_{ud}^*} \right|^2 = \tan^4 \theta_C, \tag{6.31}$$

where θ_C is the Cabibbo angle.

The time dependent decay rate of WS decays relative to RS decays can be obtained from Eq. (3.36). Assuming the oscillation effects in the RS mode to be negligible because of the small x and y parameters one obtains

$$
\begin{aligned}
R(t) &= \frac{d\Gamma(D^0 \to \bar{f})(t)}{d\Gamma(D^0 \to f)(t)} \\
&= \frac{1}{2|A_f|^2}\left[\left(|A_{\bar{f}}|^2 + \left|\bar{A}_{\bar{f}}\frac{q}{p}\right|^2\right)\cosh\frac{\Delta\Gamma}{2}t + \left(|A_{\bar{f}}|^2 - \left|\bar{A}_{\bar{f}}\frac{q}{p}\right|^2\right)\cos\Delta m\, t\right. \\
&\quad \left. -2\mathrm{Re}\left(A^*_{\bar{f}}\bar{A}_{\bar{f}}\frac{q}{p}\right)\sinh\frac{\Delta\Gamma}{2}t - 2\mathrm{Im}\left(A^*_{\bar{f}}\bar{A}_{\bar{f}}\frac{q}{p}\right)\sin\Delta m\, t\right] \\
&= \frac{R_D}{2}\left[\left(1 + \left|\frac{\bar{A}_{\bar{f}}}{A_{\bar{f}}}\frac{q}{p}\right|^2\right)\cosh\frac{\Delta\Gamma}{2}t + \left(1 - \left|\frac{\bar{A}_{\bar{f}}}{A_{\bar{f}}}\frac{q}{p}\right|^2\right)\cos\Delta m\, t\right. \\
&\quad \left. -2\mathrm{Re}\left(\frac{\bar{A}_{\bar{f}}}{A_{\bar{f}}}\frac{q}{p}\right)\sinh\frac{\Delta\Gamma}{2}t - 2\mathrm{Im}\left(\frac{\bar{A}_{\bar{f}}}{A_{\bar{f}}}\frac{q}{p}\right)\sin\Delta m\, t\right].
\end{aligned}
\tag{6.32}
$$

Under the assumption of no *CP* violation and with the dimensionless mixing parameters x and y and the the definition of the phase δ,

$$\delta = \arg\left(\frac{A_{\bar{f}}}{\bar{A}_{\bar{f}}}\frac{p}{q}\right), \tag{6.33}$$

the equation simplifies to

$$
\begin{aligned}
R(t) = \frac{R_D}{2}&\left[\left(1 + \frac{1}{R_D}\right)\cosh y\frac{t}{\tau} + \left(1 - \frac{1}{R_D}\right)\cos x\frac{t}{\tau}\right. \\
&\left. + \frac{2}{\sqrt{R_D}}\cos\delta\sinh y\frac{t}{\tau} + \frac{2}{\sqrt{R_D}}\sin\delta\sin x\frac{t}{\tau}\right],
\end{aligned}
\tag{6.34}
$$

where $\tau = 1/\Gamma$ is the D^0 lifetime. Because the D^0 mixing is a small effect, one can approximate $xt/\tau \ll 1$ and $yt/\tau \ll 1$ leading to

$$
\begin{aligned}
R(t) &= \frac{R_D}{2}\left[1 + \frac{1}{R_D} + \frac{y^2}{2R_D}\left(\frac{t}{\tau}\right)^2 + 1 - \frac{1}{R_D} + \frac{x^2}{2R_D}\left(\frac{t}{\tau}\right)^2\right. \\
&\quad \left. + \frac{2}{\sqrt{R_D}}\cos\delta\, y\frac{t}{\tau} + \frac{2}{\sqrt{R_D}}\sin\delta\, x\frac{t}{\tau}\right]. \\
&= R_D + \sqrt{R_D}\, y'\frac{t}{\tau} + \frac{x'^2 + y'^2}{4}\left(\frac{t}{\tau}\right)^2
\end{aligned}
\tag{6.35}
$$

with

$$x' = -x\,\cos\delta + y\,\sin\delta, \quad y' = x\,\sin\delta + y\,\cos\delta. \tag{6.36}$$

In case of no D^0 mixing, both parameters, x' and y', are zero and the relative decay rate to WS final states is constant in time.

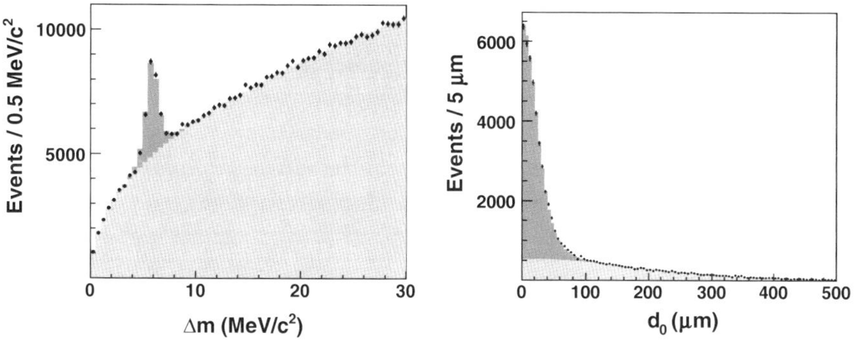

Fig. 6.15 Wrong sign $D^0 \to K^+\pi^-$ yield in bins of δm (*left*) and impact parameter distribution of D^{*+} mesons with right sign D^0 decay in the decay time bin $5 < t/\tau < 6$ (*right*) [25]. The dark shaded area shows the correctly tagged events in the left plot and the prompt events in the right plot

The decays to a pair of a kaon and pion that are used in the CDF analysis were collected by the Two-Track-Trigger. Since it is a fully reconstructed decay, the decay time is measured via the decay length with respect to the primary vertex and the momentum like in other time-dependent analyses (see. Eq. (5.2)). The production flavor is determined by requiring the D^0 to come from a $D^{*+} \to D^0\pi^+$ decay. The charge of the pion tags the flavor unambiguously.

Both decay hypotheses, RS and WS, are considered for the selected D^0 daughter tracks. To reduce the background from RS decays in the WS sample, a WS candidate is rejected if the invariant mass of the kaon and pion pair, $m_{K\pi}$, calculated for the RS hypothesis, is consistent with a RS decay. This requirement removes 96.4 % of this background and has a WS signal efficiency of 78 %. A further background reduction is achieved by using the specific ionization in the drift chamber to identify kaons and pions.

To measure the yields of WS to RS decays as a function of time, the data is split into 20 bins in t for both samples. The bin sizes are chosen to have similar yields in all bins. In each t bin the data is again split into bins in mass difference between the D^{*+} and D^0 candidates, $\delta m = m_{K\pi\pi} - m_{K\pi}$, and then the $m_{K\pi}$ distribution is fitted to separate correctly reconstructed D^0 mesons from combinatorial background and misassigned D^0 decays. The yield of D^0 mesons as a function of δm, as shown for the time-integrated WS sample in Fig. 6.15, is then fitted to obtain the number D^0 mesons that are correctly tagged by the D^{*+} decay. After these fits the WS and RS yields in bins of time are known.

A complication in this analysis arises from the fact that the sample does not only contain promptly produced D^{*+} mesons, but also D^{*+} mesons from b hadron decays. Since the decay length is calculated with respect to the primary vertex, the decay time measurement is biased for these events. To correct for the fraction of non-prompt events, the impact parameter distribution of the D^{*+} is exploited (see Fig. 6.15). While prompt events have a narrow distribution that is independent of

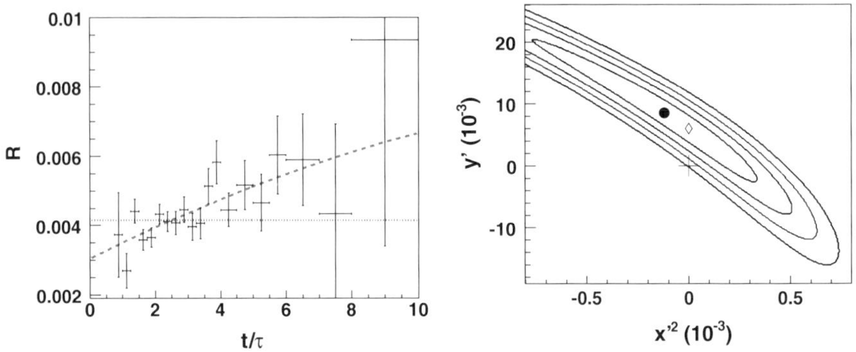

Fig. 6.16 Time dependent ratio of prompt WS and RS D^0 decays with fit result (*left*) and 1σ, 2σ, 3σ and 4σ credibility contours in the $x'^2 - y'^2$ plane (*right*) [25]. The dashed line in the left plot and the circle in the right plot show the fit result. The dotted line and cross represent the hypothesis of no mixing. The diamond shows the fit result with the constraint $x'^2 \geq 0$

the decay time, the distribution of non-prompt events is broader and becomes wider with increasing decay time.

The time-dependent ratio of corrected yields is shown in Fig. 6.16. The error bars include the systematic uncertainties which are mainly coming from the background shapes in $m_{K\pi}$, δm, and the impact parameter. One can see that the ratio rises with time as expected in case of D^0 oscillations. A fit with the formula in Eq. (6.35) yields $R_D = (3.04 \pm 0.55) \times 10^{-3}$, $y' = (8.5 \pm 7.6) \times 10^{-3}$, and $x'^2 = (-0.12 \pm 0.35) \times 10^{-3}$ [25]. Although both mixing parameters alone are consistent with zero, the strong correlation between them leads to an exclusion of the no-mixing hypothesis at the 3.8σ level. This is illustrated by the credibility regions in the right plot of Fig. 6.16. It is also obvious from the projection of the fit assuming no mixing (see the left plot in Fig. 6.16) that it does not describe the data well. The R_D value of $(4.15 \pm 0.10) \times 10^{-3}$ obtained under this assumption is consistent with the value obtained in an earlier CDF measurement of the time-integrated ratio [32], suggesting that this measurement was already affected by the D^0 oscillations.

References

1. G.A. Giurgiu, FERMILAB-THESIS-2005-41, UMI-31-86027 (2005)
2. V.M. Abazov et al., Phys. Rev. **D74**, 112002 (2006)
3. V. Tiwari, FERMILAB-THESIS-2007-09 (2007)
4. C. Lecci, FERMILAB-THESIS-2005-89, IEKP-KA-2005-13 (2005)
5. M. Feindt, U. Kerzel, Nucl. Instrum. Meth. **A559**, 190 (2006)
6. A. Abulencia et al., Phys. Rev. Lett. **97**, 242003 (2006)
7. T. Aaltonen et al., Phys. Rev. **D85**, 072002 (2012)
8. A. Abulencia et al., Phys. Rev. Lett. **97**, 062003 (2006)
9. J. Charles et al., Eur. Phys. J. **C41**, 1 (2005)

10. M. Bona et al., JHEP **0507**, 028 (2005)
11. H. Moser, A. Roussarie, Nucl. Instrum. Meth. **A384**, 491 (1997)
12. J. Abdallah et al., Eur. Phys. J. **C35**, 35 (2004)
13. A. Heister et al., Eur. Phys. J. **C29**, 143 (2003)
14. G. Abbiendi et al., Eur. Phys. J. **C11**, 587 (1999)
15. K. Abe et al., Phys. Rev. **D67**, 012006 (2003)
16. V.M. Abazov et al., Phys. Rev. Lett. **97**, 021802 (2006)
17. A. Buras, W. Slominski, H. Steger, Nucl. Phys. **B245**, 369 (1984)
18. A.J. Buras, M. Jamin, P.H. Weisz, Nucl. Phys. **B347**, 491 (1990)
19. T. Inami, C. Lim, Prog. Theor. Phys. **65**, 297 (1981)
20. M. Okamoto, PoS LAT2005, 013 (2006).
21. J. Beringer et al., Phys. Rev. D **86**, 010001 (2012)
22. R. Aaij et al., Phys. Lett. **B709**, 177 (2012)
23. M. Staric et al., Phys. Rev. Lett. **98**, 211803 (2007)
24. B. Aubert et al., Phys. Rev. Lett. **98**, 211802 (2007)
25. T. Aaltonen et al., Phys. Rev. Lett. **100**, 121802 (2008)
26. S.L. Glashow, J. Iliopoulos, L. Maiani, Phys. Rev. **D2**, 1285 (1970)
27. L. Wolfenstein, Phys. Lett. B**164**, 170 (1985)
28. J.F. Donoghue, E. Golowich, B.R. Holstein, J. Trampetic, Phys. Rev. **D33**, 179 (1986)
29. H. Georgi, Phys. Lett. B**297**, 353 (1992)
30. A.F. Falk, Y. Grossman, Z. Ligeti, A.A. Petrov, Phys. Rev. **D65**, 054034 (2002)
31. A.F. Falk, Y. Grossman, Z. Ligeti, Y. Nir, A.A. Petrov, Phys. Rev. **D69**, 114021 (2004)
32. A. Abulencia et al., Phys. Rev. **D74**, 031109 (2006)

Chapter 7
CP-Violation in Mixing and the Interference of Mixing and Decay

Because the measured B_s^0 mixing frequency agrees well with the standard model expectation, a contribution from new physics processes to the magnitude of the mixing amplitude must be small. But it is still possible that the phase of the mixing amplitude deviates significantly from its standard model value. The "golden mode" to search for such a new physics effect is $B_s^0 \to J/\psi\phi$.

7.1 $B_s^0 \to J/\psi\phi$

A B_s^0 meson can decay either directly to the $J/\psi\phi$ final state via a tree level process, or via an oscillation to a $\bar{B}_s^0$ meson followed by a tree level decay of the $\bar{B}_s^0$. The Feynman diagrams of both processes are shown in Fig. 7.1. The interference between these processes can lead to mixing-induced *CP* violation. Because the decay to $J/\psi\phi$ is dominated by a single amplitude, direct *CP* violation is expected to be negligible.

The magnitude of the mixing-induced *CP* violation is determined by the phase between the two diagrams, $\phi_s^{J/\psi\phi}$. Neglecting higher order loop processes, it is given in the standard model by

$$\phi_s^{J/\psi\phi,SM} = -2\beta_s := -2\arg\left(-\frac{V_{ts}V_{tb}^*}{V_{cs}V_{cb}^*}\right). \tag{7.1}$$

Here and in the following discussion we assume no *CP* violation in mixing and in the decay as both are expected to be negligible. In the standard model also the mixing induced *CP* violation is small with $-2\beta_s \approx -0.04$ [1]. This makes it a good probe for new physics because any sizable deviation from zero would be an indication of contributions from processes beyond the standard model.

If new physics processes would contribute to the mixing loop diagram they would change the phase ϕ_s between the off-diagonal elements of the mass and decay matrices as defined in Eq. (3.19) from its standard model value, ϕ_s^{SM}, by a term ϕ_s^{NP}:

T. Kuhr, *Flavor Physics at the Tevatron*, Springer Tracts in Modern Physics 249,
DOI: 10.1007/978-3-642-10300-1_7,

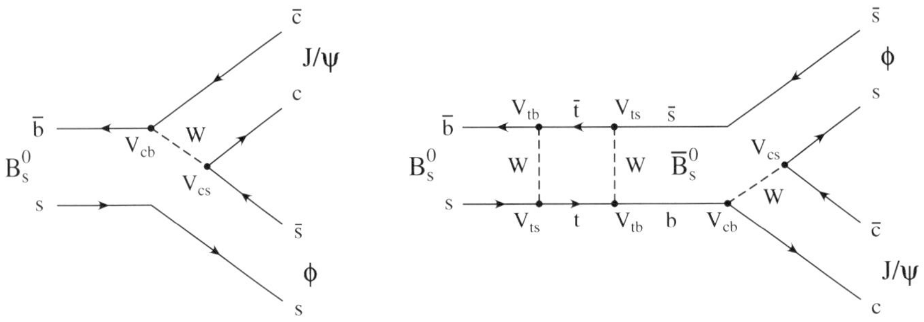

Fig. 7.1 Feynman diagrams of $B_s^0 \to J/\psi\phi$ decays without and with mixing

$$\phi_s = \phi_s^{SM} + \phi_s^{NP}. \tag{7.2}$$

This phase shift would also show up in the mixing-induced *CP* violation in $B_s^0 \to J/\psi\phi$ decays:

$$\phi_s^{J/\psi\phi} = -2\beta_s + \phi_s^{NP}. \tag{7.3}$$

Because β_s and $\phi_s^{SM} = 4 \times 10^{-3}$ [1] are both much smaller than the current experimental sensitivity, one can set them approximately to zero and obtain the relation

$$\phi_s^{J/\psi\phi} = \phi_s^{NP} = \phi_s. \tag{7.4}$$

Different naming conventions are used in the literature. While D0 quotes its results in terms of ϕ_s, CDF prefers to use $-2\beta_s^{J/\psi\phi}$, or just $-2\beta_s$, for the same quantity. So one has to keep in mind a sign flip and a factor of two when comparing results from both experiments. In the following we will call the measured *CP* violating phase ϕ_s.

As discussed in Sect. 3.2, *CP* violation in the interference between mixing and decay can be observed in a decay to a final state that is accessible by B_s^0 and $\bar{B}_s^0$ mesons. If the final state is a *CP* eigenstate and the decay width difference, $\Delta\Gamma_s$, is neglected, the time evolution of B_s^0 mesons is simply given by an exponential decay modulated by an oscillation with frequency Δm_s and an amplitude that is equal to the sinus of the *CP* violating phase (cf. Eq. (3.80)).

For the decay $B_s^0 \to J/\psi\phi$ the situation is more complicated because the decay width difference has to be taken into account and the final state is not a pure *CP* eigenstate. To understand the time-dependent $B_s^0 \to J/\psi\phi$ decay rate, we separate the *CP*-even and *CP*-odd components of the final state:

$$|f\rangle = |J/\psi\phi\rangle = |f_{even}\rangle + |f_{odd}\rangle \tag{7.5}$$

with

$$CP\ |f_{even}\rangle = |f_{even}\rangle \quad \text{and} \quad CP\ |f_{odd}\rangle = -|f_{odd}\rangle\,. \tag{7.6}$$

The decay amplitude as defined in Eq. (3.35) can then be split in two terms as well:

$$A_f = \langle f|B_s^0\rangle = A_{even} + A_{odd} \tag{7.7}$$

with

$$A_{even} = \langle f_{even}|B_s^0\rangle \quad \text{and} \quad A_{odd} = \langle f_{odd}|B_s^0\rangle. \tag{7.8}$$

For the decay amplitude of $\bar{B}_s^0$ mesons follows:

$$\bar{A}_f = \langle f|\bar{B}_s^0\rangle = A_{even} - A_{odd}. \tag{7.9}$$

With the definition of the strong phase difference between A_{even} and A_{odd} as

$$\delta = \arg\left(A_{even}^* A_{odd}\right) \tag{7.10}$$

the following expressions can be calculated:

$$\begin{aligned} |A_f|^2 &= (A_{even}^* + A_{odd}^*)(A_{even} + A_{odd}) = |A_{even}|^2 + |A_{odd}|^2 + 2\mathrm{Re}(A_{even}^* A_{odd}) \\ &= |A_{even}|^2 + |A_{odd}|^2 + 2|A_{even}||A_{odd}|\cos\delta && (7.11) \\ |\bar{A}_f|^2 &= |A_{even}|^2 + |A_{odd}|^2 - 2\mathrm{Re}(A_{even}^* A_{odd}) \\ &= |A_{even}|^2 + |A_{odd}|^2 - 2|A_{even}||A_{odd}|\cos\delta && (7.12) \\ A_f^* \bar{A}_f &= |A_{even}|^2 - |A_{odd}|^2 - 2i\mathrm{Im}(A_{even}^* A_{odd}) \\ &= |A_{even}|^2 - |A_{odd}|^2 - 2i|A_{even}||A_{odd}|\sin\delta. && (7.13) \end{aligned}$$

Inserting these terms and $\phi_s = \arg(q/p)$ in Eq. (3.36) yields

$$\begin{aligned} &\frac{d\Gamma[B_s^0 \to J/\psi\phi](t)}{dt} \\ &= N_f e^{-\Gamma t}\left\{\left(|A_{even}|^2 + |A_{odd}|^2\right)\cosh\frac{\Delta\Gamma}{2}t\right. \\ &\quad + 2|A_{even}||A_{odd}|\cos\delta\cos\Delta m t \\ &\quad - \left[(|A_{even}|^2 - |A_{odd}|^2)\cos\phi_s + 2|A_{even}||A_{odd}|\sin\delta\sin\phi_s\right]\sinh\frac{\Delta\Gamma}{2}t \\ &\quad \left. - \left[(|A_{even}|^2 - |A_{odd}|^2)\sin\phi_s + 2|A_{even}||A_{odd}|\sin\delta\cos\phi_s\right]\sin\Delta m t\right\} \\ &= N_f\left[|A_{even}|^2 T_+(t) + |A_{odd}|^2 T_-(t)\right. \\ &\quad \left. + 2|A_{even}||A_{odd}|U(t)\right] && (7.14) \end{aligned}$$

with

$$\begin{aligned} T_{\pm}(t) &= e^{-\Gamma t}\left[\cosh\frac{\Delta\Gamma}{2}t \mp \cos\phi_s \sinh\frac{\Delta\Gamma}{2}t \mp \xi \sin\phi_s \sin\Delta m\, t\right] \\ &= \frac{1\mp\cos\phi_s}{2}e^{-\Gamma_H t} + \frac{1\pm\cos\phi_s}{2}e^{-\Gamma_L t} \mp \xi \sin\phi_s e^{-\Gamma t}\sin\Delta m\, t \end{aligned} \tag{7.15}$$

$$\begin{aligned} U(t) = e^{-\Gamma t}\Big[&\sin\delta\sin\phi_s\sinh\frac{\Delta\Gamma}{2}t \\ &+ \xi(\sin\delta\cos\phi_s\sin\Delta m\, t - \cos\delta\cos\Delta m\, t)\Big], \end{aligned} \tag{7.16}$$

and $\xi = +1$. The decay rate for $\bar{B}_s^0$ mesons is obtained by setting ξ to -1. The terms $T_{\pm}(t)$ describe the time evolution of the pure *CP* eigenstate components. The interference between both components gives rise to the term $U(t)$.

In case the production flavor is not known and the initial number of B_s^0 and $\bar{B}_s^0$ mesons is the same, as can be safely assumed for B_s^0 mesons produced via strong interation in $p\bar{p}$ collisions, the time evolution of the untagged mesons is described by

$$\begin{aligned} T_{\pm}^{\text{untagged}}(t) &= e^{-\Gamma t}\left[\cosh\frac{\Delta\Gamma}{2}t \mp \cos\phi_s\sinh\frac{\Delta\Gamma}{2}t\right] \\ &= \frac{1\mp\cos\phi_s}{2}e^{-\Gamma_H t} + \frac{1\pm\cos\phi_s}{2}e^{-\Gamma_L t} \end{aligned} \tag{7.17}$$

$$U^{\text{untagged}}(t) = e^{-\Gamma t}\sin\delta\sin\phi_s\sinh\frac{\Delta\Gamma}{2}t. \tag{7.18}$$

Several interesting features of untagged decays can be deduced from the above relations:

- In case of $\Delta\Gamma = 0$ the interference term U vanishes and the T terms become simple exponential functions, $e^{-\Gamma t}$. All terms containing ϕ_s disappear.
- If $\Delta\Gamma \neq 0$, the T terms are the sum of two exponentials with different decay rates. The relative fractions depend on ϕ_s and are opposite for the two *CP* final states. This provides sensitivity to the *CP* violating phase if the lifetime difference can be resolved and the *CP* final states can be distinguished experimentally. For $\phi_s = 0$ the *CP*-even component would be $e^{-\Gamma_L t}$ and the *CP*-odd component would be $e^{-\Gamma_H t}$. Thus in case of *CP* conservation the B_s^0 eigenstate with decay width Γ_L (Γ_H) would always decay to a *CP*-even (*CP*-odd) final state. A deviation from this behavior would be a sign of *CP* violation. This way of measuring *CP* violation is the same as was used for the discovery of *CP* violation in the kaon system.
- In case of $\Delta\Gamma \neq 0$ the interference term provides additional sensitivity to ϕ_s if there is a strong phase difference between A_{even} and A_{odd} so that $\sin\delta \neq 0$.
- There are two symmetries. Equations (7.17) and (7.18) are invariant under the transformations

$$\Delta\Gamma \to -\Delta\Gamma, \quad \phi_s \to \pi - \phi_s, \quad \delta \to -\delta \tag{7.19}$$

$$\phi_s \to -\phi_s, \quad \delta \to -\delta. \tag{7.20}$$

This leads to a four-fold ambiguity and makes an untagged measurement insensitive to the signs of $\Delta\Gamma$ and ϕ_s.

When the information about the initial flavor of the B_s^0 meson is added, the time evolution is described by Eqs. (7.15) and (7.16) and the following advantages are obtained:

- The term $\xi \sin\phi_s \sin\Delta m\, t$ in Eq. (7.15) provides additional sensitivity to the *CP* violating phase, ϕ_s. It is independent of strong phases and observable in the decays of the pure *CP* components. It gives rise to the time-dependent asymmetry defined in Eq. (3.80). This way of measuring mixing-induced *CP* violation is the same as was used for the discovery of *CP* violation in the B^0 system.
- A further term containing ϕ_s also enters in the interference component in Eq. (7.16) if $\sin\delta \neq 0$.
- Sensitivity to Δm is provided by the term $\cos\delta \cos\Delta m\, t$ in Eq. (7.16), independently of the *CP*-violating phase.
- The second symmetry of the untagged decay rate in Eq. (7.20) is resolved. However, the first one in Eq. (7.19) remains. Thus the tagging reduced the four-fold to a two-fold ambiguity.

As the above discussion shows, one of the experimental challenges is to identify the *CP* eigenvalue of the final state. The mixture of *CP* eigenstates in the $J/\psi\phi$ final state is caused by the fact that both daughter particles are vector mesons. Their spins can be combined to a total spin of 0, 1, or 2. To conserve angular momentum in the decay of the pseudoscalar B_s^0 meson, the total spin has to be compensated by the orbital angular momentum, L. Since J/ψ and ϕ have both even *CP* parity, the total *CP* eigenvalue is given by $(-1)^L$. Thus the $J/\psi\phi$ final state is *CP*-even for $L = 0$ and $L = 2$, and *CP*-odd for $L = 1$.

The different orbital angular momenta lead to different angular distributions of the final state particles. Thus the identification of the *CP* eigenvalue of the final state requires an angular analysis. Instead of the orbital angular momenta basis, the transversity basis [2] is commonly used to describe the $J/\psi\phi$ final state. The transversity basis defines two angles in the J/ψ rest frame where the x-axis is given by the momentum direction of the B_s^0 meson and the xy-plane is defined by the $\phi \to K^+K^-$ decay plane with the y component of the K^+ momentum being positive. The angles θ_T and ϕ_T are then the polar and azimuthal angles of the μ^+ momentum, respectively. The third angle, ψ_T, is the ϕ helicity angle, defined as the angle between the K^+ and the negative B_s^0 momentum in the ϕ rest frame. Figure 7.2 illustrates the definition of the angles in the transversity basis. In the following we will denote them $\rho = (\cos\theta_T, \phi_T, \cos\psi_T)$.

As there are three different spin states, the decay is described by three amplitudes. The common choice of amplitude definition refers to the linear polarization states of the vector mesons. The longitudinal component is given by A_0. For transverse

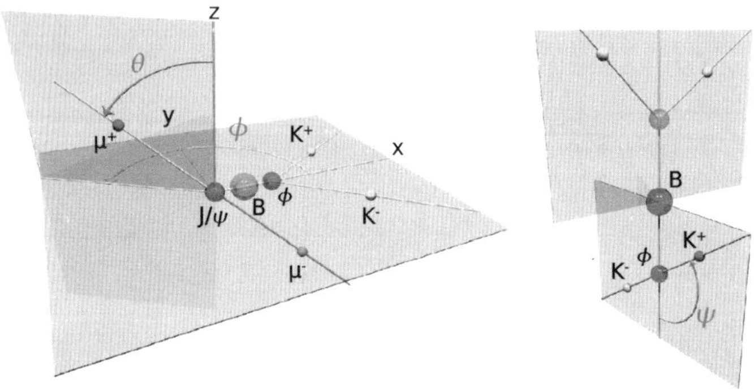

Fig. 7.2 Definition of angles θ, ϕ, and ψ in the transversity basis

polarization, $A_{||}$ and $A_{\perp}$ describe the components with parallel and perpendicular polarization vectors, respectively. The $A_{\perp}$ amplitude corresponds to the $L = 1$ decay and therefore describes the *CP*-odd component. The *CP*-even component is given by A_0 and $A_{||}$ which are linear combinations of the *S*- and *D*-wave amplitudes. The amplitudes are defined such that they are normalized:

$$|A_0|^2 + |A_{||}|^2 + |A_{\perp}|^2 = 1. \tag{7.21}$$

Because only relative phases are relevant, the phase of A_0 is set to zero and the phases of $A_{||}$ and $A_{\perp}$ are defined with respect to A_0:

$$\delta_{||} = \arg\left(A_{||}A_0^*\right), \quad \delta_{\perp} = \arg\left(A_{\perp}A_0^*\right). \tag{7.22}$$

With these definitions of decay amplitudes at $t = 0$ the time- and angle-dependent decay rate, $d\Gamma[B_s^0 \to J/\psi\phi](t, \rho)/dt\, d\rho$, can be calculated. Explicit formulas can be found in Ref. [2], or in a more compact form in Ref. [3].

A time-integrated measurement of decay amplitudes was already performed in an angular analysis of $B_s^0 \to J/\psi\phi$ decays at CDF Run I [4]. The first time dependent measurements were done in Run II. CDF used a dataset of 260 pb^{-1} [5] and D0 a dataset of 450 pb^{-1} [6]. The main target of these analyses is the measurement of the decay width difference. It is determined in a fit to the invariant mass, decay time and angular distribution of $B_s^0 \to J/\psi\phi$ candidates under the assumption of no *CP* violation ($\phi_s = 0$). While CDF includes all three angles, D0 integrates over $\cos\theta_T$ and ϕ_T and uses only $\cos\psi_T$ which is most sensitive to the *CP*-odd fraction. The signal is described by the theoretically derived decay rate, $d\Gamma[B_s^0 \to J/\psi\phi](t, \rho)/dt\, d\rho$, convolved with a decay time resolution function and multiplied by an angle-dependent acceptance function. The sculpting of the angular distributions is caused by detector, trigger, and selection efficiencies and determined from simulation. The background consists of prompt and non-prompt candidates. While CDF assumes decay time, mass, and angles of the background to be independent, D0 allows for different mass and angular distributions for the prompt and non-prompt component.

To verify the analysis procedure, CDF performs a time-dependent angular analysis of $B^0 \to J/\psi K^{*0}$, with $K^{*0} \to K^+\pi^-$. As the K^{*0} is a vector meson, the decay is also described by three amplitudes. It is related to the $B_s^0 \to J/\psi\phi$ decay via $SU(3)$ flavor symmetry because both decay processes differ only by the flavor of the spectator quark (cf. Fig. 7.1). In case of exact flavor symmetry the amplitudes of both decays are expected to be identical.

In contrast to $B_s^0 \to J/\psi\phi$, the $B^0 \to J/\psi K^{*0}$ decay is flavor specific, with the flavor being tagged by the charge of the kaon. In the reconstruction of the K^{*0} the assignment of the kaon and pion mass hypotheses to the two daughter tracks is not always perfect. The fraction of signal events with swapped assignment is taken into account in the fit. The amplitudes, phases, and B^0 lifetime determined in the fit to the $B^0 \to J/\psi K^{*0}$ sample are consistent with previous measurements.

The first measurements of the decay width difference in the B_s^0 meson system are $\Delta\Gamma/\Gamma = (0.65^{+0.25}_{-33}\ \text{(stat)} \pm 0.01\ \text{(syst)})$ by CDF [5] and $\Delta\Gamma/\Gamma = 0.24^{+0.28}_{-0.38}\ \text{(stat)}^{+0.03}_{-0.04}\ \text{(syst)}$ by D0 [6]. Both measurements could not yet establish a finite decay width difference. CDF observes agreement between the B_s^0 and B^0 amplitudes, suggesting that $SU(3)$ flavor symmetry holds at the tested level of precision. This is confirmed by a time-dependent angular analysis of $B_s^0 \to J/\psi\phi$ and $B^0 \to J/\psi K^{*0}$ decays by D0 with a dataset of 2.8 fb^{-1} [7]. The $\Delta\Gamma$ value measured in this analysis is $(0.085^{+0.075}_{-0.078}\ \text{(stat)} \pm 0.006\ \text{(syst)})\ \text{ps}^{-1}$.

An analysis which included for the first time the effect of CP violation was carried out by D0 on a data sample of 1.1 fb^{-1} [8] and by CDF on a data sample of 1.7 fb^{-1} [9]. In this analysis, D0 uses all three angles and measures $\phi_s = -079 \pm 0.56\ \text{(stat)}^{+0.14}_{-0.01}\ \text{(syst)}$. The decay width difference is measured as well and the likelihood contour plot in Fig. 7.3 left shows the correlation between ϕ_s and $\Delta\Gamma$. In the untagged analysis, CDF observes a fit bias when the CP violation is small as it is the case in the standard model. The bias can be understood from Eq. (7.18). For $\phi_s = 0$ the term $U^{\text{untagged}}(t)$ vanishes and the differential decay rate becomes insensitive to the strong phase difference δ. The effective loss of a degree of freedom leads to a bias away from the standard model values. Figure 7.4 illustrates this effect on pseudoexperiments. To take this bias into account, CDF presents the result as a confidence region, shown in Fig. 7.3 right, that uses the likelihood-ratio ordering method of Feldman and Cousins [10]. The p-value that quantifies the agreement of the data with the standard model hypothesis is determined to be 22%.

Both result plots show the four-fold ambiguity discussed above. For $\phi_s = \pm\pi/2$, the B_s^0 mass and lifetime eigenstates are mixtures of the CP eigenstates with equal fraction so that the angular analysis cannot separate them any more and the analysis becomes less sensitive to $\Delta\Gamma$. The shaded bands for the new physics models are obtained from Eq. (3.18)

$$\Delta\Gamma \approx 2|\Gamma_{12}| \cos\phi_s \approx \Delta\Gamma_{SM} \cos\phi_s, \tag{7.23}$$

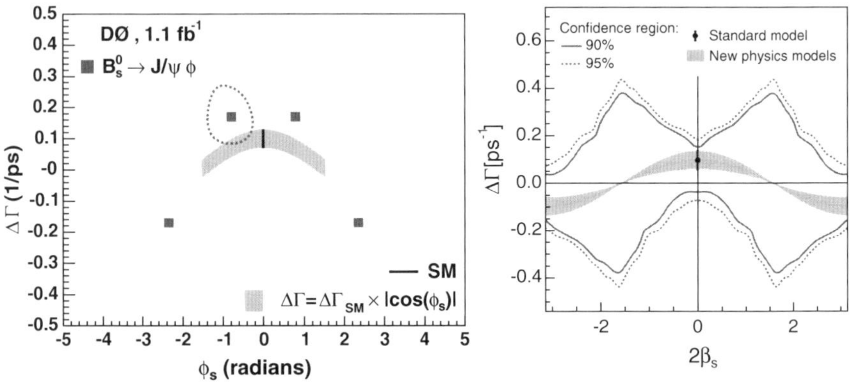

Fig. 7.3 Four solutions and $\Delta\mathscr{L} = 0.5$ contour for one of the solutions as dotted line in the $\phi_s - \Delta\Gamma$ plane measured by D0 [8] (left) and confidence regions in the $\beta_s - \Delta\Gamma$ plane measured by CDF [9] (*right*). The *black line* at $\phi_s = -2\beta_s \approx 0$ shows the standard model prediction

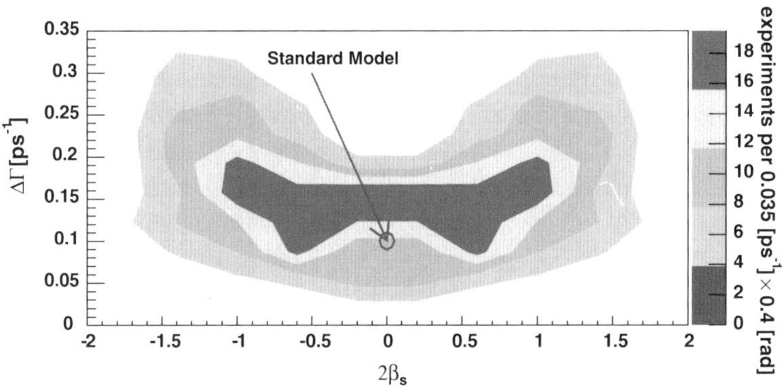

Fig. 7.4 Distribution of fitted $(\beta_s, \Delta\Gamma)$ values for pseudoexperiments generated with the standard model values as input [11]

where $\Delta\Gamma_{SM}$ is the decay width difference in the standard model. Since Γ_{12} is dominated by tree level decays like $B_s^0 \to D_s^+ D_s^-$ (see Sect. 5.3), new physics contributions are assumed to be negligible. A recent theoretical calculation yields $\Delta\Gamma_{SM} = (0.087 \pm 0.021)\ \text{ps}^{-1}$ [12]. New physics contributions are expected to only have sizable effects on ϕ_s. Such contributions are not yet significantly constrained by the untagged measurements.

To improve the sensitivity to the *CP*-violating phase ϕ_s, both collaborations added tagging information in the analyses. The first tagged analysis by CDF is based on 1.3 fb^{-1} [13], the first D0 analysis uses 2.8 fb^{-1} [14]. Because of the limited sensitivity to the mixing frequency, Δm_s is fixed (D0) or constrained (CDF) to the world average value. In addition to the tagging algorithms described in Sect. 6.1, D0 also employs a same side kaon tagger in this analysis.

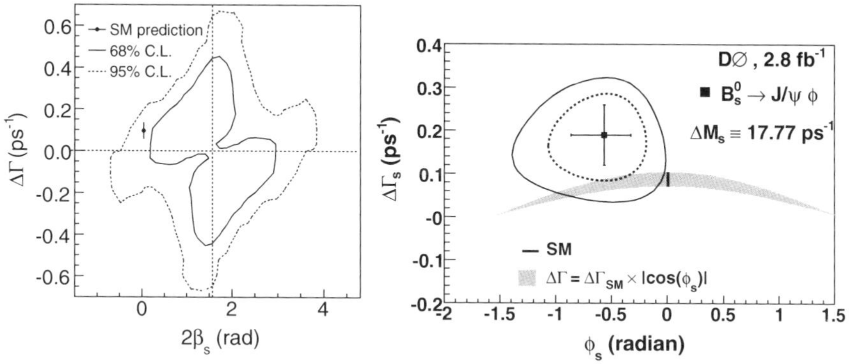

Fig. 7.5 First results in the $\phi_s - \Delta\Gamma$ plane of tagged $B_s^0 \to J/\psi\phi$ analyses by CDF [13] (*left*) and D0 [14] (*right*)

The results are shown in Fig. 7.5. Systematic uncertainties have been determined to be negligible. One can see in the CDF plot that the tagging has resolved one symmetry and only a two-fold ambiguity remains. The latter ambiguity is resolved by D0 by assuming $SU(3)$ flavor symmetry and constraining the strong phases to the world average values of the $B^0 \to J/\psi K^{*0}$ decay. Keeping in mind that $\phi_s = -2\beta_s$, both measurements are in agreement and seem to prefer a negative ϕ_s value. The standard model p-values are quoted as 15% and 6.6% by CDF and D0, respectively.

The results seemed to support models that predict a large CP-violating phase in the B_s^0 system. But also the concern was raised, that the results might be biased due to the neglection of a contribution from $B_s^0 \to J/\psi K^+K^-$ decays where the kaon pair is non-resonant or from a $f_0(980) \to K^+K^-$ decay [15]. Since the non-resonant component is expected to have mainly spin 0, both contributions together are referred to as S-wave. A sizable S-wave contribution and in particular its interference with the P-wave ϕ decay would lead to modified angular distributions. While such an S-wave contribution would be an issue for the measurements presented above, it would have the advantage that it provides the possibility to resolve the remaining ambiguity. The method to resolve the ambiguity exploits the fact that the phase of the P-wave component varies strongly around the ϕ pole mass while the the S-wave phase is nearly constant over the range where both interfere. One of the two solutions could be chosen based on a measurement of the phase difference as a function of the K^+K^- invariant mass. It is challenging because it requires a good mass resolution, but it would make assumptions about $SU(3)$ flavor symmetry obsolete. This kind of measurement was already used to resolve the ambiguity in the mixing-induced CP-violation measurement in the B^0 system [16].

CDF and D0 addressed the issue of the S-wave component by including it in the signal PDF in updated analyses of 5.2 fb^{-1} [17] and 8 fb^{-1} of data [18], respectively. CDF uses a neural network for the signal selection and optimizes the requirement on its output with pseudoexperiments to minimize the expected ϕ_s uncertainty. Such a procedure is also applied by D0 for a boosted decision tree based selection. In addition

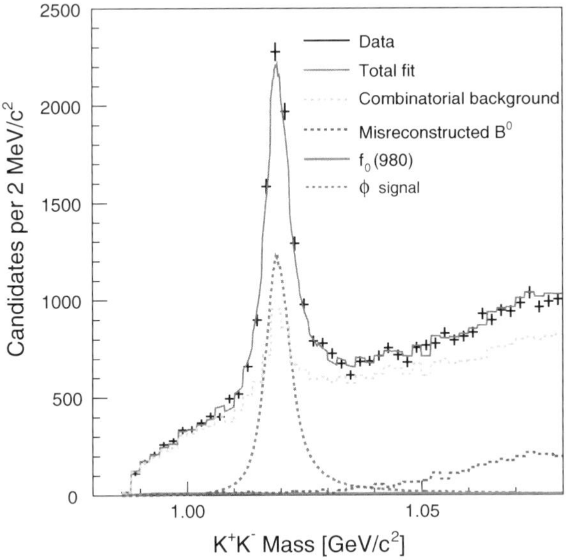

Fig. 7.6 Fit of the kaon pair invariant mass distribution including a possible S-wave contribution denoted as $f_0(980)$ [17]

D0 performs the analysis with an alternative selection using simple requirements on kinematic and quality variables.

Because of the limited mass resolution and the difficulty to include the kaon pair mass in the background PDFs no attempt is made to measure the decay rate differentially in $m(K^+K^-)$. Instead the signal distribution is integrated over $m(K^+K^-)$ in the selected range assuming a flat S-wave component and a Breit-Wigner model for the P-wave, where CDF takes the relativistic, and D0 the non-relativistic version. The relative fraction of the S-wave component is given by the parameter F_S. A detailed definition and formulas for the differential decay rate including the S-wave can be found in Ref. [3]. CDF measures an S-wave fraction consistent with zero and quotes an upper limit of $F_S < 6\%$ at 95% confidence level for $1.009 < m(K^+K^-)/\text{GeV} < 1.028$ [17]. The result of the time-dependent angular analysis is supported by the kaon pair invariant mass distribution, shown in Fig. 7.6, which is well described without any S-wave component. This is in contradiction with the D0 result of $F_S = (17.3 \pm 3.6)\%$ for $1.01 < m(K^+K^-)/\text{GeV} < 1.003$ [18]. From a comparison of B_s^0 yields in the ϕ signal region and sideband, D0 estimates an S-wave fraction of $(12 \pm 3)\%$. Such a large value would be welcome for the resolution of the ambiguity, but a measurement by LHCb favors a small S-wave contribution [19].

To obtain the result on the CP-violating phase ϕ_s and the decay width difference $\Delta\Gamma$, CDF determines contours in the 2-dimensional profile likelihood distribution. To account for deviations from the ideal Gaussian likelihood shape, a coverage adjust-

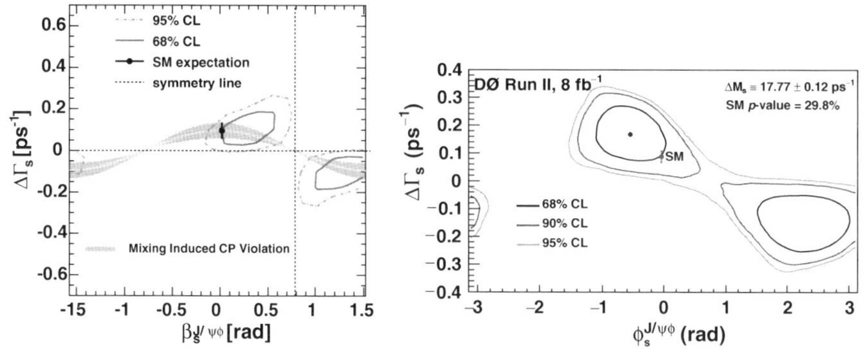

Fig. 7.7 Results in the $\phi_s - \Delta\Gamma$ plane of tagged $B_s^0 \to J/\psi\phi$ analyses including an S-wave component by CDF [17] (*left*) and D0 [18] (*right*). In the D0 result the constraint $\cos\delta_\perp < 0$ is applied

ment is applied. The change in the likelihood function value corresponding to a 68 and 95% confidence level is determined with different sets of pseudoexperiments generated at randomly chosen values of the nuisance parameters and the standard model values of ϕ_s and $\Delta\Gamma$. This coverage adjustment factor in the likelihood value is shown to be independent of the $(\phi_s, \Delta\Gamma)$ value and therefore applied for all points in the 2D plane in the determination of the confidence region. This method avoids that pseudoexperiments have to be generated at each $(\phi_s, \Delta\Gamma)$ point. The same approach is used to obtain 1-dimensional confidence intervals on ϕ_s. D0 determines credibility regions and point estimates with Markov Chain Monte Carlos (MCMC) [20]. This Bayesian method is also used to combine the results obtained with the two different selections after verifying that they are consistent with each other.

The results in the $\phi_s - \Delta\Gamma$ plane are shown in Fig. 7.7. Compared with its previous measurement, CDF could considerably reduce the allowed values of ϕ_s and $\Delta\Gamma$. The data is now in good agreement with the standard model hypothesis. The p-value is 30%. For the 1-dimensional 68% confidence interval CDF quotes $\phi_s \in [-3.10, -2.16] \cup [-1.04, -0.04]$ [17]. D0 shows only one of the two equivalent solutions by imposing the requirement $\cos\delta_\perp < 0$ which selects the solution preferred by $SU(3)$ flavor symmetry. The fact that there are still two minima of the likelihood function is caused by the additional symmetry in the untagged case. Because only opposite side taggers are used here, the tagging power is not sufficient to resolve the ambiguity at a significant level. The standard model p-value is 30%. As a point estimate, D0 quotes $\phi_s = -0.56^{+0.38}_{-0.36}$ [18].

D0 also measures the decay width difference, the mean lifetime, $\tau = 1/\Gamma$, and the linear polarization amplitudes with the MCMC method where systematic uncertainties are included. CDF determines these parameters in a likelihood fit under the assumption of the standard model value for ϕ_s. The results are summarized in Table 7.1. Systematic uncertainties on the angular acceptance, the PDF models,

Table 7.1 Measured parameters of $B_s^0 \to J/\psi\phi$ decays by CDF [17] and D0 [18]

Parameter	CDF	D0
$\Delta\Gamma$ [ps^{-1}]	0.075 ± 0.035 (stat) ± 0.006 (syst)	$0.163^{+0.065}_{-0.064}$
τ [ps]	1.529 ± 0.025 (stat) ± 0.012 (syst)	$1.443^{+0.038}_{-0.035}$
$\lvert A_0\rvert^2$	0.524 ± 0.013 (stat) ± 0.015 (syst)	$0.558^{+0.017}_{-0.019}$
$\lvert A_{\parallel}\rvert^2$	0.231 ± 0.014 (stat) ± 0.015 (syst)	$0.231^{+0.024}_{-0.030}$

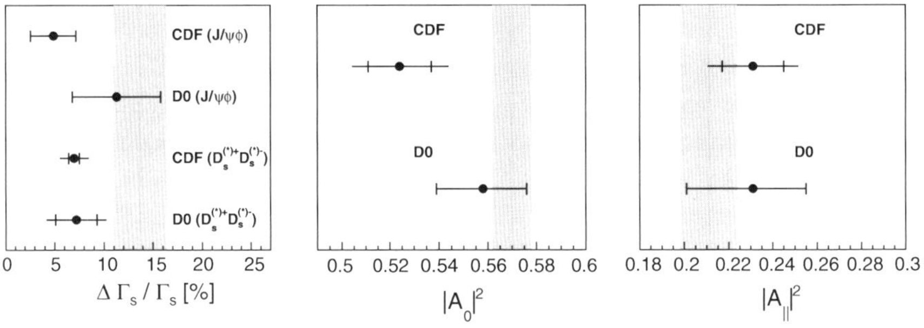

Fig. 7.8 Results of the tagged $B_s^0 \to J/\psi\phi$ analyses by CDF [17] and D0 [18] compared with other measurements and predictions. The *left plot* shows the measured decay with difference in comparison with the $\Delta\Gamma$ values obtained from $\mathscr{B}(B_s^0 \to D_s^{(*)+}D_s^{(*)-})$ measurements [21, 22] and the standard model expectation [12], shown as band. The *middle* and *right plots* show comparisons of the measured polarizations with the world average values for $B^0 \to J/\psi K^{*0}$ decays [23]

$B^0 \to J/\psi K^{*0}$ cross feed where the pion from the K^{*0} is misidentified as a kaon, and the vertex detector alignment are considered.

A comparison of the lifetime measurements with other measurements were already shown in Fig. 5.24 in Sect. 5.4. In Fig. 7.8, the $\Delta\Gamma$ value are compared with the theoretical prediction and the $\Delta\Gamma$ value obtained from $\mathscr{B}(B_s^0 \to D_s^{(*)+}D_s^{(*)-})$ measurements under the assumption that these decays saturate $\Delta\Gamma$ as discussed in Sect. 5.3. One can see that the measurements agree with the standard model expectation and the $\Delta\Gamma$ values calculated from the $B_s^0 \to D_s^{(*)+}D_s^{(*)-}$ branching ratios. So the experimental data does not yet tell whether there is a significant contribution to $\Delta\Gamma$ from decays other than $B_s^0 \to D_s^{(*)+}D_s^{(*)-}$. The middle and right plots in Fig. 7.8 show that the polarization amplitudes are consistent with the ones measured in $B^0 \to J/\psi K^{*0}$ decays as expected in the case of $SU(3)$ flavor symmetry.

7.2 $B_s^0 \to D_s^-\mu^+X$

While the $B_s^0 \to J/\psi\phi$ analysis is sensitive to CP violation in the interference of decays with and without mixing, a further way to search for new physics processes is to measure CP violation in B_s^0 mixing via an asymmetry in flavor specific decay rates.

The semileptonic asymmetry for B_s^0 mesons, a_{SL}^s, defined according to Eq. (3.61) , depends on the phase ϕ_s via the relation

$$a_{SL}^s = -\frac{|\Gamma_{12}|}{|M_{12}|} \sin \phi_s = -\frac{\Delta\Gamma}{\Delta m_s} \tan \phi_s \tag{7.24}$$

as shown in Eqs. (3.64) and (3.65) . In the standard model a_{SL}^s is expected to be tiny, $(1.9 \pm 0.3) \times 10^{-5}$ [12]. This is well below the experimental sensitivity so that a measurement that deviates significantly from zero can be regarded as a sign of new physics.

Assuming $|p/q| \approx 1$, Eq. (3.62) can be used to derive

$$1 + a_{SL}^s = \frac{2\left|\frac{p}{q}\right|^2}{\left|\frac{p}{q}\right|^2 + \left|\frac{q}{p}\right|^2} \approx \left|\frac{p}{q}\right|^2 \tag{7.25}$$

$$1 - a_{SL}^s = \frac{2\left|\frac{q}{p}\right|^2}{\left|\frac{p}{q}\right|^2 + \left|\frac{q}{p}\right|^2} \approx \left|\frac{q}{p}\right|^2 . \tag{7.26}$$

In case of no CP violation in the decay, the time-dependent decay rates of B_s^0 and $\bar{B}_s^0$ mesons to a flavor specific final state f_{fs}, as given in Eqs. (3.43)–(3.46), can then be expressed with a_{SL}^s:

$$\frac{d\Gamma[B_s^0 \to f_{fs}](t)}{dt} = \frac{1}{2} N_f |A_f|^2 e^{-\Gamma t} \left[\cosh \frac{\Delta\Gamma}{2} t + \cos \Delta m_s\, t\right] \tag{7.27}$$

$$\frac{d\Gamma[\bar{B}_s^0 \to f_{fs}](t)}{dt} = \frac{1}{2} N_f |A_f|^2 (1 + a_{SL}^s) e^{-\Gamma t} \left[\cosh \frac{\Delta\Gamma}{2} t - \cos \Delta m_s\, t\right] \tag{7.28}$$

$$\frac{d\Gamma[B_s^0 \to \bar{f}_{fs}](t)}{dt} = \frac{1}{2} N_f |A_f|^2 (1 - a_{SL}^s) e^{-\Gamma t} \left[\cosh \frac{\Delta\Gamma}{2} t - \cos \Delta m_s\, t\right] \tag{7.29}$$

$$\frac{d\Gamma[\bar{B}_s^0 \to \bar{f}_{fs}](t)}{dt} = \frac{1}{2} N_f |A_f|^2 e^{-\Gamma t} \left[\cosh \frac{\Delta\Gamma}{2} t + \cos \Delta m_s\, t\right] . \tag{7.30}$$

In case the initial flavor of the B_s^0 is not tagged, only the asymmetry between the decay rates to the two flavor specific final states can be observed

$$a_{untagged}^s = \frac{d\Gamma[B_s^0/\bar{B}_s^0 \to f_{fs}] - d\Gamma[B_s^0/\bar{B}_s^0 \to \bar{f}_{fs}]}{d\Gamma[B_s^0/\bar{B}_s^0 \to f_{fs}] + d\Gamma[B_s^0/\bar{B}_s^0 \to \bar{f}_{fs}]} = \frac{1}{2} a_{SL}^s (1 - \cos \Delta m_s\, t). \tag{7.31}$$

Because of the fast B_s^0 oscillation the cos term vanishes in a time-integrated measurement and the following relation is obtained:

$$a_{SL}^s = 2a_{untagged}^s. \tag{7.32}$$

Such a time-integrated untagged measurement was first done by D0 with a dataset of 1.3 fb^{-1} [24]. B_s^0 mesons are partially reconstructed in the decay $B_s^0 \to D_s^- \mu^+ X$ with $D_s^- \to \phi\pi^-$ and $\phi \to K^+K^-$. The selection is optimized with a likelihood ratio method and the signal yield is extracted from a fit to the $\phi\pi^-$ invariant mass distribution.

To measure $a_{untagged}^s$, the sample is first split into events with positive and negative muon charge. The asymmetry between the yields of $D_s^-\mu^+$ and $D_s^+\mu^-$ events is not only affected by the *CP* violation in B_s^0 mixing. A pseudorapidity-dependent asymmetry can be caused by a forward-backward asymmetry of $b\bar{b}$ production in $p\bar{p}$ collisions. To take this into account the sample is further split by the sign of the pseudorapidity of the $D_s\mu$ candidate. Finally, the detection efficiency can depend on the charge, the pseudorapidity, and the polarity of the magnetic field which is reverted regularly in the D0 detector. By splitting the sample also in magnet polarity and fitting the yields in all eight subsamples, the physics and instrumental asymmetries can be determined.

The physics asymmetries measured for background events are consistent with zero as expected. The composition of B decays contributing to the $D_s\mu$ signal yield and thus to the measured asymmetry is taken from simulation. The contribution from events where D_s and μ come from different c or b quarks is measured on data. With a fraction of $(83.2 \pm 3.3)\%$ of the $D_s\mu$ events coming from semileptonic B_s^0 decays an asymmetry for these signal events of $a_{untagged}^s = [12.3 \pm 9.7\,(\text{stat}) \pm 1.7\,(\text{syst})] \times 10^{-3}$ corresponding to $a_{SL}^s = [24.5 \pm 19.3\,(\text{stat}) \pm 3.5\,(\text{syst})] \times 10^{-3}$ is measured [24] where the dominant systematic uncertainty comes from the fitting procedure.

This analysis was refined by D0 on a larger dataset of 5 fb^{-1} [25]. In this analysis the $D_s^- \to K^{*0}K^-$ decay mode with $K^{*0} \to K^+\pi^-$ is added and a time-dependent measurement is performed where the initial B_s^0 flavor is determined by the opposite side tagger. The invariant mass spectra of $D_s\mu$ candidates for both D_s decay modes are shown in Fig. 7.9. The detector induced muon reconstruction asymmetry is measured with $J/\psi \to \mu^+\mu^-$ events.

The result is $a_{SL}^s = [-1.7 \pm 9.1\,(\text{stat})^{+1.4}_{-1.5}\,(\text{syst})] \times 10^{-3}$ [25], well consistent with the standard model expectation. The largest systematic uncertainty comes from the kaon detection asymmetry in the $D_s^- \to K^{*0}K^-$ mode, but the total systematic uncertainty is still much smaller than the statistical one. Compared to the previous result of the untagged analysis the uncertainty is reduced by a factor two. Since the data sample is increased by about a factor of four, the improvement can be mainly attributed to the increase in statistics. With the partially reconstructed decays the fast B_s^0 oscillations cannot be resolved well so that the cos terms in Eqs. (7.27)–(7.30) taken into account in the tagged analysis provide little additional sensitivity.

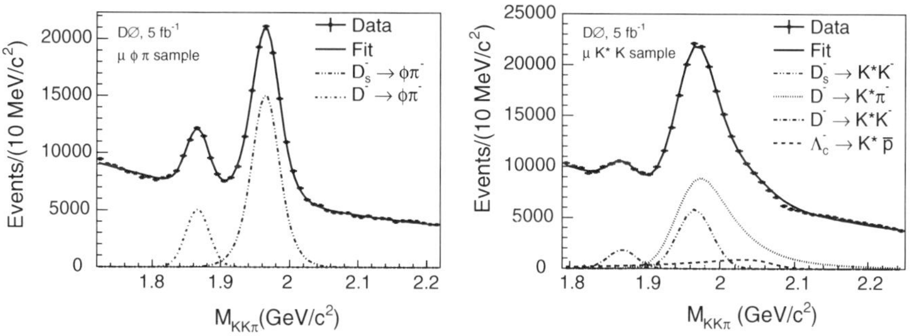

Fig. 7.9 Invariant mass distribution of $D_s^- \to \phi\pi^-$ (*left*) and $\phi \to K^+K^-$ candidates (*right*) [25]

7.3 Dimuon Charge Asymmetry

While the reconstruction of $D_s\mu$ pairs in the analysis described in the previous Section enriches the fraction of B_s^0 mesons in the selected sample, only the charge of the muon is required to tag its flavor at decay time. A more inclusive approach is thus the measurement of the asymmetry of positively and negatively charged muons. However, the contributions from semileptonic B^0 and B_s^0 decays cannot be distinguished experimentally any more. Moreover, several background sources lead to a strong dilution of the measured asymmetry. This can be improved by requiring that the other b-hadron produced in the $b\bar{b}$ production process decays semileptonically, too. The second muon does not only reduce the background, it also provides a flavor tag. In events where both muons from the direct semileptonic decays of the two b-hadrons have the same charge one of the hadrons must have decayed in the oscillated and the other in the non-oscillated state. Because *CP* violation in mixing is observable in the "wrong-sign" decays the asymmetry between positively and negatively charged dimuon decay rates is therefore well suited to measure a_{SL}.

The dimuon charge asymmetry for B mesons is defined by the time integrated decay rates as

$$A_{SL}^b = \frac{\Gamma[b\bar{b} \to \mu^+\mu^+X] - \Gamma[b\bar{b} \to \mu^-\mu^-X]}{\Gamma[b\bar{b} \to \mu^+\mu^+X] + \Gamma[b\bar{b} \to \mu^-\mu^-X]}, \tag{7.33}$$

where only direct b hadron decays are considered and sequential $b \to c \to \mu$ decays are regarded background processes. Since the "wrong-sign" decays are only possible for B^0 and B_s^0 mesons and not for B^+ mesons and b baryons follows

$$\Gamma[b\bar{b} \to \mu^+\mu^+X] = \Gamma[\bar{B}_{(s)}^0 \to \mu^+X]\Gamma[\bar{b} \to \mu^+X] \tag{7.34}$$

$$\Gamma[b\bar{b} \to \mu^-\mu^-X] = \Gamma[B_{(s)}^0 \to \mu^-X]\Gamma[b \to \mu^-X]. \tag{7.35}$$

In case of no direct *CP* violation the "right-sign" decay rates, $\Gamma[\bar{b} \to \mu^+ X]$ and $\Gamma[b \to \mu^- X]$, are equal as one can see in Eq. (7.27) and (7.30). So they cancel in A^b_{SL} and the asymmetry between "wrong-sign" decays remains. The dimuon charge asymmetry in events with B^0_s mesons is then obtained with Eq. (7.28) and (7.29) as $A^s_{SL} = a^s_{SL}$. The same relation holds for B^0 mesons, $A^d_{SL} = a^d_{SL}$. The experimentally observed dimuon charge asymmetry is a combination of A^d_{SL} and A^s_{SL},

$$A^b_{SL} = \beta_d A^d_{SL} + (1 - \beta_d) A^s_{SL}, \tag{7.36}$$

where β_d is the fraction of the B^0 contribution. It depends on the production fractions, f_d and f_s, and fraction of "wrong-sign" decays, χ_d and χ_s, of B^0 and B^0_s mesons, respectively:

$$\beta_d = \frac{f_d \chi_d}{f_d \chi_d + f_s \chi_s}. \tag{7.37}$$

A first measurement of the dimuon charge asymmetry by D0 was based on a data set of 1.0 fb^{-1} [26]. The experimental observable is the asymmetry in the yields of like-sign dimuon events with positive and negative charge, N^{++} and N^{--}, respectively:

$$A = \frac{N^{++} - N^{--}}{N^{++} + N^{--}}. \tag{7.38}$$

The effect of different reconstruction efficiencies for μ^+ and μ^- is taken into account by averaging the yields for data taking periods with opposite magnet polarities. This exploits the fact that the trajectory of a μ^+ is identical to that of a μ^- with reverted field. Any remaining detection asymmetry is assumed to be negligible and included in the systematics.

Besides the direct semileptonic decays of *b*-hadron pairs, which yield 64 % of the dimuon events, several background processes can contribute to *A*. These include sequential $b \to c \to \mu$ decays, muons from prompt charm hadrons, $c\bar{c}$ or $b\bar{b}$ resonances, or Drell-Yan processes. Also cosmic rays, muons from kaon decays, and hadrons misidentified as muons are considered. The relative fractions of these contributions are determined with simulation, data from cosmic ray runs, and measured misidentification rates of kaons in $D^{*+} \to D^0(\to K^-\pi^+)\pi^+$ events.

The only background with a significant asymmetry are muons from kaon decays combined with a prompt muon. The asymmetry is a result of the different interaction cross sections of K^+ and K^- with the detector material. After correcting for this effect a value of $A^b_{SL}/\beta_d = [-9.2 \pm 4.4\,(\text{stat}) \pm 3.2\,(\text{syst})] \times 10^{-3}$ is obtained [26], where the dominant systematic uncertainty comes from the correction of the kaon background. Using Eq. (7.24) this result and the a^s_{SL} measurement with untagged $B^0_s \to D^-_s \mu^+ X$ decays [24] is combined with the untagged $B^0_s \to J/\psi\phi$ measurement by D0 [8] to obtain $\phi_s = -0.70^{+0.47}_{-0.39}$ [27].

A considerable improvement in the analysis technique was introduced by D0 in an updated measurement using 6.1 fb^{-1} [28, 29]. The main idea is to determine in

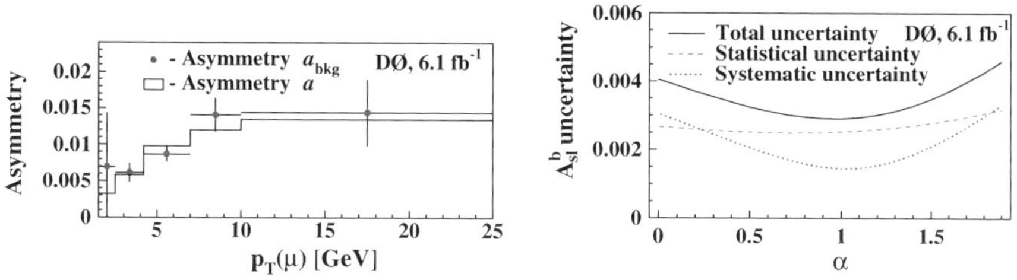

Fig. 7.10 Cross-check of the muon misidentification fractions and asymmetries (*left*) and dependency of the A^b_{SL} uncertainty on α [29]

addition to A the asymmetry of single muon event yields with positive and negative charge, n^+ and n^-, respectively,

$$a = \frac{n^+ - n^-}{n^+ + n^-}, \tag{7.39}$$

and combine both asymmetries so that systematic uncertainties are reduced. Also the treatment of backgrounds and detection asymmetries are refined.

The fraction of background events with a hadron misidentified as a muon, including decay in flight, is measured on data in bins of transverse momentum of the muon track candidate. The decay $K^{*0} \to K^+\pi^-$ is used for kaons with a normalization taken from $K^{*-} \to K^0_S\pi^-$ and exploiting isospin symmetry. The fraction of misidentified pions and protons is obtained from the misidentification probability relative to that of kaons, measured with $K^0_S \to \pi^+\pi^-$, $\Lambda \to p\pi^-$, and $\phi \to K^+K^-$ events, and the relative particle multiplicities taken from simulation. The above mentioned decay modes are also used to determine the asymmetries of the backgrounds. They are calculated in bins of p_T from the difference and sum of event yields with positive and negative track charge. While the measured asymmetries for kaons are of the order of 4–6 %, the values for pions are below 2 %. The proton measurements are consistent with zero within uncertainties of a few percent. To verify the measured fractions and asymmetries of misidentified hadrons, these numbers are used to calculate expected a values in the single muon sample where contributions from other processes to the asymmetry are negligible. As can be seen in the left plot of Fig. 7.10, the calculation agrees well with the measured asymmetries.

After the subtraction of backgrounds from misidentified hadrons, only events with real muons are left. The detection asymmetry, that is remaining after averaging over magnet polarities, is measured with $J/\psi \to \mu^+\mu^-$ events. The p_T-dependent values are in the range of ± 0.2 %.

The observed asymmetries after correcting for the detection asymmetry have contributions from several physics processes which are studied in detail with simulation. The only processes exhibiting a difference in the yield of μ^+ and μ^- are oscillated B^0_s and B^0 mesons decaying either directly to a muon or via a flavor-specific sequential decay of a charm hadron to a muon. Because the muon from the charm decay has opposite sign, both contribute with opposite sign to the asymmetry. All other considered processes have no asymmetry so that they just dilute the measured value. The factor by which the asymmetry is scaled is determined to be 0.070 ± 0.006 for the single muon sample and 0.486 ± 0.032 for the dimuon sample. The values illustrate that the single muon sample is dominated by background and thus less sensitive to A^b_{SL}.

The measured asymmetries for B mesons are $A^b_{SL} = [+9.4 \pm 11.2\,(\text{stat}) \pm 21.4\,(\text{syst})] \times 10^{-3}$ for the single muon sample and $A^b_{SL} = [-7.36 \pm 2.66\,(\text{stat}) \pm 3.05\,(\text{syst})] \times 10^{-3}$ for the like-sign dimuon sample, where the dominant systematic uncertainties come from the fraction of misidentified kaons. The statistical errors include the statistical uncertainties of the quantities measured on data, namely the background fractions and asymmetries and the detection asymmetries. Both results are consistent within uncertainties and could be averaged. Instead, the fact that both numbers are affected by backgrounds in a similar way can be exploited by considering the observable $A - \alpha a$. As shown in the right plot of Fig. 7.10, the largest reduction of uncertainties is obtained for $\alpha \approx 1$ and the measured asymmetry for the optimal value of $\alpha = 0.959$ is $A^b_{SL} = [-9.57 \pm 2.51\,(\text{stat}) \pm 1.46\,(\text{syst})] \times 10^{-3}$ [28, 29]. Several cross-checks with alternative event selections are performed to verify the stability of the result.

The result caused some excitement, because it is 3.2 standard deviations away from the standard model expectation of $A^b_{SL}(SM) = [-0.28^{+0.06}_{-0.05}] \times 10^{-3}$, calculated from Eq. (7.36) with predictions of a^d_{SL} and a^s_{SL} [1] and $\beta_d = 0.594 \pm 0.022$ [30]. Since the source of the muons is not explicitly identified one cannot say whether the deviation, if not a statistical fluctuation, comes from the *CP* violation in the mixing of B^0_s or B^0 mesons, or both, or some other process.

An attempt to identify the reason for the deviation from the standard model expectation was made in an improved analysis by D0 using 9.0 fb^{-1} [30]. One of the improvements is an increase in the muon acceptance by 25 % due to a softer requirement on the momentum of the muon tracks. About half of the gain in statistics remains after applying tighter requirements on the matching of tracks in the central tracker and the muon detector which reduces the background from decay in flight hadrons.

The fraction of kaons misidentified as muons in the single muon sample is determined more accurately by requiring one of the pions from the K^0_S of the $K^{*-} \to K^0_S \pi^-$ normalization mode to be identified as a muon instead of requiring an additional muon in the event to avoid a bias in the sample composition. The uncertainty due to the fraction of misidentified kaons in the like-sign dimuon sample is reduced by averaging the values of two independent measurements. The first method uses $K^{*0} \to K^+\pi^-$ decays as in the previous analysis, but refines the method by fitting a combination of the $K^+\pi^-$ invariant mass distributions for single and dimuon samples so that uncertainties in mass resolution and backgrounds largely cancel. The second method measures the ratio of kaons in the dimuon to single muon sample with $K^0_S \to \pi^+\pi^-$ decays, assuming isospin invariance. A small correction is applied for the correlation between the charges of the misidentified kaon and the second

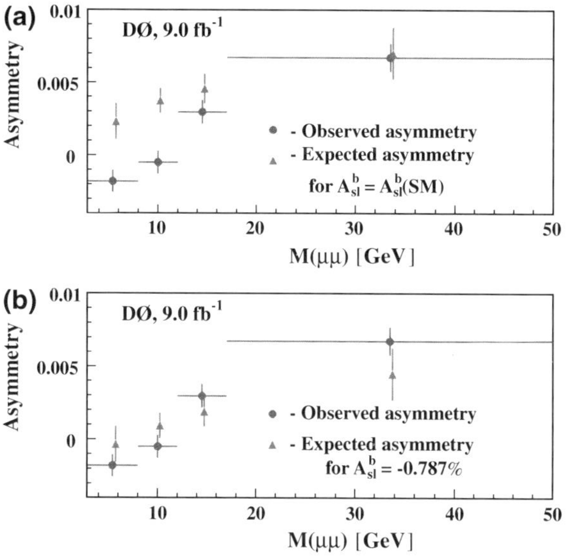

Fig. 7.11 Measured and expected A distribution for $A^b_{SL} = A^b_{SL}(SM)$ (*top*) and $A^b_{SL} = -7.87 \times 10^{-3}$ (*bottom*) [30]

muon. The measurements of the fractions agree within uncertainties and half of the difference is taken as systematic uncertainty.

Using otherwise the same method as in the previous analysis, a value of $A^b_{SL} = [-10.4 \pm 13.0\,(\text{stat}) \pm 23.1\,(\text{syst})] \times 10^{-3}$ for the single muon sample and $A^b_{SL} = [-8.08 \pm 2.02\,(\text{stat}) \pm 2.22\,(\text{syst})] \times 10^{-3}$ for the like-sign dimuon sample is measured. The combined result, treating all uncertainties except for the statistical uncertainties of a, A, and the relative fraction of misidentified kaons as fully correlated, is $A^b_{SL} = [-7.87 \pm 1.72\,(\text{stat}) \pm 0.93\,(\text{syst})] \times 10^{-3}$ [30]. As a cross-check of the result Fig. 7.11 shows the observed value of A as a function of dimuon invariant mass compared with the expectation for an A^b_{SL} value equal to the standard model prediction and the measured value. Only in the latter case the data is well described, indicating that the physics processes and backgrounds which have different dependencies of their asymmetry on $M(\mu\mu)$ are modeled correctly.

The A^b_{SL} result disagrees with the standard model prediction at a level of $3.9\,\sigma$. This is illustrated in the left plot of Fig. 7.12 where the relation between a^d_{SL} and a^s_{SL} is taken from Eq. (7.36). The result is consistent with direct measurements, but their precision does not allow to decide whether the discrepancy with respect to the standard model prediction is driven by a^d_{SL} or a^s_{SL}.

Information about the origin of the large A^b_{SL} value is obtained by splitting the sample in events with a muon impact parameter below and above 120 μm. While the fraction of oscillated B^0_s mesons is the same in both samples because of the fast B^0_s oscillation, the fraction of oscillated B^0 mesons is larger in the second sample

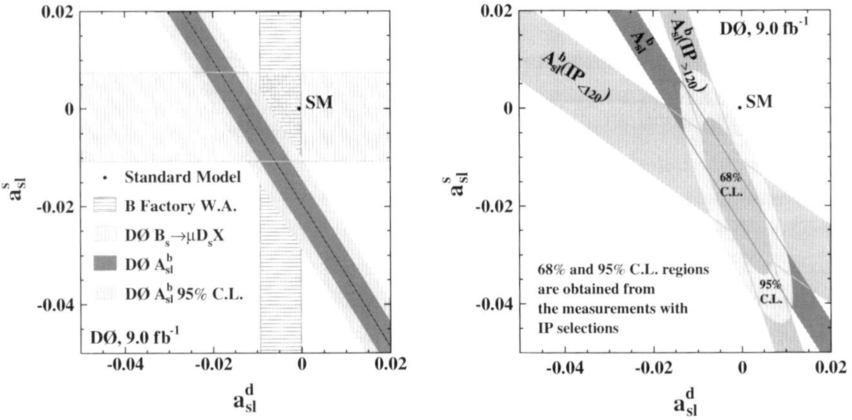

Fig. 7.12 Constraint of the A^b_{SL} measurements in the a^s_{SL} versus a^d_{SL} plane compared with the standard model expectation [30]. The *left* plot also shows direct measurements [25, 31], and the *right* plot shows constraints obtained from samples with different muon impact parameter requirements

because the B^0 oscillation period is of the same order as the B^0 decay time. This leads to an increased β_d value of 0.729 ± 0.018 in the sample with larger impact parameter, and $\beta_d = 0.397 \pm 0.022$ in the other sample. A large impact parameter requirement also reduces the background in the single muon sample considerably so that in the combination $A - \alpha a$ a negative value of α yields the best result. Because the measured asymmetries of $A^b_{SL} = [-5.79 \pm 2.10\,(\text{stat}) \pm 0.94\,(\text{syst})] \times 10^{-3}$ for the high impact parameter sample and $A^b_{SL} = [-11.4 \pm 3.7\,(\text{stat}) \pm 3.2\,(\text{syst})] \times 10^{-3}$ for the low impact parameter sample are linear combinations of a^d_{SL} or a^s_{SL} (see Eq. (7.36)) with different values of β_d, one can determine the individual semileptonic asymmetries of B^0 and B^0_s mesons. The result is $a^d_{SL} = (1.2 \pm 5.2) \times 10^{-3}$ and $a^s_{SL} = (-18.1 \pm 10.5) \times 10^{-3}$ with a correlation of $-80\,\%$ [30]. A graphical representation of the measurement is shown in the right plot of Fig. 7.12. Although the measured asymmetry for B^0_s mesons is much larger than that of B^0 mesons, the large uncertainties do not allow to draw a definite conclusion about the origin of the large like-sign dimuon charge asymmetry.

The A^b_{SL} result is compared with the values obtained in the predecessor dimuon analyses in the left plot of Fig. 7.13. All values are below the standard model expectation and one can see that not only the statistical uncertainties are reduced with increasing data size, but also the systematic uncertainties get smaller with improved analysis techniques. The right plot in Fig. 7.13 compares the a^s_{SL} result obtained from the dimuon charge asymmetry measurements with different muon impact parameter requirements with the a^s_{SL} determination from $B^0_s \to D^-_s \mu^+ X$ decays described in the previous Section. Both results are consistent with each other and the standard model prediction. However, the central value of the standard model prediction is about three orders of magnitude smaller than the measurements. To improve the precision of the a^s_{SL} determination one can consider external information for a^d_{SL} in

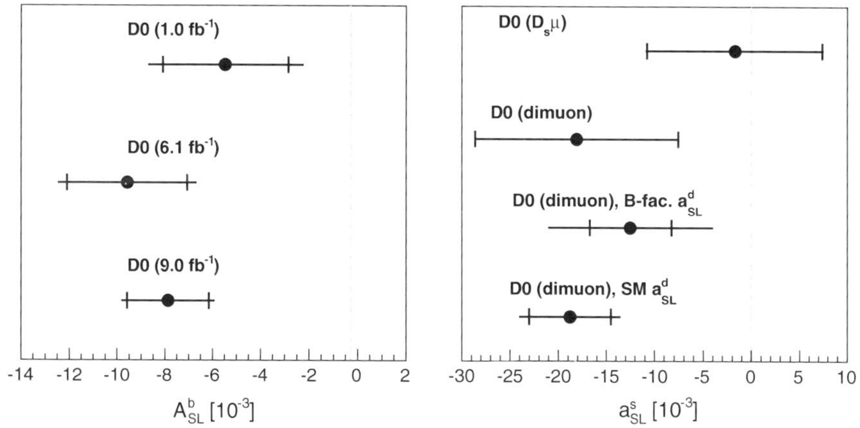

Fig. 7.13 Comparison of the A^b_{SL} measurements with like-sign dimuon events [26, 28–30] (*left*) and a^s_{SL} results obtained from $B^0_s \to D^-_s \mu^+ X$ decays [25] and the dimuon charge asymmetry [30] (*right*). Also shown are a^s_{SL} values calculated from A^b_{SL} with the a^d_{SL} value taken from the average of the B-factory measurements [31] and the standard model prediction [12]. The vertical bands show the standard model expectations for A^b_{SL} and a^s_{SL} [12]

the calculation of a^s_{SL} from A^b_{SL} according to Eq. (7.36). Unfortunately the precision of the measurement of a^d_{SL} at the B-factories is not sufficient to reduce the uncertainty significantly. If the standard model prediction of a^d_{SL} is taken its uncertainty is negligible and a deviation from $a^s_{SL}(SM) = (0.019 \pm 0.003) \times 10^{-3}$ [12] with a significance of the same order as in the A^b_{SL} measurement is obtained.

Looking at Eq. (7.24), which relates a^s_{SL} to $\Delta\Gamma$, Δm_s, and ϕ_s, it seems that a large asymmetry could be explained by a ϕ_s value close to $\pm\pi/2$ so that $\tan\phi_s$ becomes large. But this consideration neglects that $\Delta\Gamma = 2|\Gamma_{12}|\cos\phi_s$ (see Eq. (3.18)) and thus would require $|\Gamma_{12}|$ to become larger as well. A better relation for the discussion of possible a^s_{SL} values is obtained by expressing it in terms of Γ_{12} and Δm_s:

$$a^s_{SL} = -\frac{|\Gamma_{12}|}{2\Delta m_s}\sin\phi_s. \tag{7.40}$$

The B^0_s oscillation frequency is measured precisely and Γ_{12} can be calculated by theory. Since Γ_{12} is dominated by tree level decays a large new physics contribution to it would be observable in other measurements as well and is thus very unlikely. We can therefore take the standard model prediction of $2|\Gamma_{12}| = (0.087 \pm 0.021)\ \text{ps}^{-1}$ [12] as a reliable estimate also for the case of new physics in the B^0_s system. In combination with the measured Δm_s value one obtains from Eq. (7.40)

$$a^s_{SL} = -(4.9 \pm 1.2) \times 10^{-3}\sin\phi_s. \tag{7.41}$$

Considering further that a^s_{SL} contributes to A^b_{SL} with a weight of $1 - \beta_d = 0.406 \pm 0.022$ it is obvious that it cannot explain a central A^b_{SL} value of -7.87×10^{-3} even for maximal $\sin \phi_s$. To identify the source of the large A^b_{SL} value and to hopefully resolve the tension with theoretical models further measurements are required and expected to be delivered by LHCb.

References

1. A. Lenz, U. Nierste, JHEP **0706**, 072 (2007)
2. A.S. Dighe, I. Dunietz, R. Fleischer, Eur. Phys. J. **C6**, 647 (1999)
3. F. Azfar, J. Boudreau, N. Bousson, J. Fernandez, K. Gibson et al., JHEP **1011**, 158 (2010)
4. T. Affolder et al., Phys. Rev. Lett. **85**, 4668 (2000)
5. D.E. Acosta et al., Phys. Rev. Lett. **94**, 101803 (2005)
6. V.M. Abazov et al., Phys. Rev. Lett. **95**, 171801 (2005)
7. V.M. Abazov et al., Phys. Rev. Lett. **102**, 032001 (2009)
8. V.M. Abazov et al., Phys. Rev. Lett. **98**, 121801 (2007)
9. T. Aaltonen et al., Phys. Rev. Lett. **100**, 121803 (2008)
10. G.J. Feldman, R.D. Cousins, Phys. Rev. **D57**, 3873 (1998)
11. M. Milnik, FERMILAB-THESIS-2007-38, IEKP-KA-2007-16 (2007)
12. A. Lenz, U. Nierste, TTP11-03, TUM-HEP-792-11 (2011)
13. T. Aaltonen et al., Phys. Rev. Lett. **100**, 161802 (2008)
14. V.M. Abazov et al., Phys. Rev. Lett. **101**, 241801 (2008)
15. S. Stone, L. Zhang, Phys. Rev. **D79**, 074024 (2009)
16. B. Aubert et al., Phys. Rev. **D71**, 032005 (2005)
17. T. Aaltonen et al., Phys. Rev. **D85**, 072002 (2012)
18. V.M. Abazov et al., Phys. Rev. **D85**, 032006 (2012)
19. R. Aaij et al., Phys. Rev. Lett. **108**, 101803 (2012)
20. W. Hastings, Biometrika **57**, 97 (1970)
21. T. Aaltonen et al., Phys. Rev. Lett. **108**, 201801 (2012)
22. V.M. Abazov et al., Phys. Rev. Lett. **102**, 091801 (2009)
23. J. Beringer et al., Phys. Rev. D **86**, 010001 (2012)
24. V.M. Abazov et al., Phys. Rev. Lett. **98**, 151801 (2007)
25. V.M. Abazov et al., Phys. Rev. **D82**, 012003 (2010)
26. V.M. Abazov et al., Phys. Rev. **D74**, 092001 (2006)
27. V.M. Abazov et al., Phys. Rev. **D76**, 057101 (2007)
28. V.M. Abazov et al., Phys. Rev. Lett. **105**, 081801 (2010)
29. V.M. Abazov et al., Phys. Rev. **D82**, 032001 (2010)
30. V.M. Abazov et al., Phys. Rev. **D84**, 052007 (2011)
31. D. Asner, et al., arXiv:1010.1589 (2010)

Chapter 8
Direct *CP*-Violation

As explained in Sect. 3.2, *CP* violation in the decay requires the interference of at least two diagrams. The experimentally observable difference in decay rates of *CP* conjugated modes, $A_{CP} := \mathscr{A}_{decay}$ as defined in Eq. (3.50), is according to Eq. (3.55) proportional to $|A_1||A_2| \sin \Delta\delta \sin \Delta\phi$, where $|A_1|$ and $|A_2|$ are the magnitudes of the two decay amplitudes and $\Delta\delta$ and $\Delta\phi$ the difference of their strong and weak phases, respectively. While $\Delta\phi$ can usually be calculated precisely, predictions of $\Delta\delta$ have often large uncertainties which makes the theoretical interpretation of A_{CP} measurements difficult. Although direct *CP* violation measurements are therefore sometimes considered less attractive than indirect *CP* violation measurements, results like the difference in direct *CP* violation in B^+ and B^0 decays to $K\pi$ [1] or $D^0 \to K^+K^-$ and $D^0 \to \pi^+\pi^-$ [2] have shown unexpectedly large values and caused excitement and speculations about first hints of new physics contributions. The contributions from the Tevatron to the direct *CP* violation measurements will be dicussed in this Chapter, starting with $b \to c\bar{c}s$ and $b \to s\bar{s}s$ transitions, followed by $B \to DK$ decays which are sensitive to the angle γ of the unitarity triangle, and finally covering charm meson decays where hints for new physics might emerge.

8.1 $B^+ \to J/\psi K^+$

The decay $B^+ \to J/\psi K^+$ is dominated by the Cabibbo-favored tree level $b \to c\bar{c}s$ transition. Thus the direct *CP* violation in this mode is expected to be small. The A_{CP} value is predicted to be about 0.003 in the standard model and up to the order of 0.01 in case of new physics [3]. A somewhat larger direct *CP* violation effect of the order of a percent is expected in the decay $B^+ \to J/\psi\pi^+$ because the $b \to c\bar{c}d$ transition is Cabibbo-suppressed.

The decay rate asymmetry between B^- and B^+ mesons was measured by D0 for both modes with a dataset of 2.8 fb^{-1} [4]. Signal candidates are selected by a likelihood ratio method and the signal yields are determined in a fit to the $J/\psi K$ invariant

T. Kuhr, *Flavor Physics at the Tevatron*, Springer Tracts in Modern Physics 249,
DOI: 10.1007/978-3-642-10300-1_8, © Springer-Verlag Berlin Heidelberg 2013

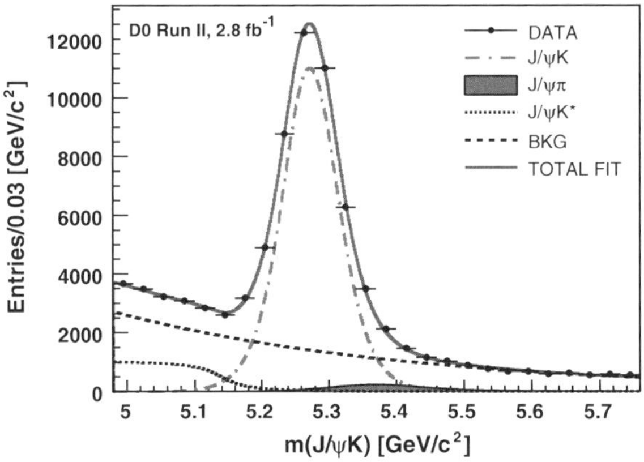

Fig. 8.1 Invariant mass distribution of $B^+ \to J/\psi K^+$ candidates with fit result [4]

mass distribution. In addition to the two signal modes and combinatorial background, a contribution from $B \to J/\psi K^*$ where the pion from the K^* decay is not reconstructed is considered. A dependency of the mass shapes on the kaon momentum is taken into account in the fit model. A projection of the fit is shown in Fig. 8.1.

To disentangle the effects of direct *CP* violation, a forward-backward asymmetry of $b\bar{b}$ production in $p\bar{p}$ collisions, and detection asymmetries, the sample is split by magnet polarity, sign of the pseudorapidity of the *B* candidate, and charge of the kaon. Using the same method as in the like-sign dimuon charge asymmetry measurement described in Sect. 7.3, the $J/\psi K$ and $J/\psi\pi$ asymmetries are determined from the fitted yields in the subsamples. These asymmetries still have to be corrected for the difference in interaction rate of K^+ and K^- with the detector material. The correction is measured with $D^{*+} \to D^0[\to \mu^+\nu_\mu K^-]\pi^+$ events as a function of the kaon momentum. The results are $A_{CP}(B^+ \to J/\psi K^+) = +0.0075 \pm 0.0061$ (stat) $\pm$ 0.0027 (syst) and $A_{CP}(B^+ \to J/\psi\pi^+) = -0.09 \pm 0.08$ (stat) $\pm$ 0.03 (syst) [4], where the dominant systematic uncertainties come from the mass model. Both values are consistent with zero and agree with measurements by BaBar and Belle as one can see in Fig. 8.2.

8.2 Charmless Hadronic *B* Decays

As large direct *CP* violation can only arise when the amplitudes of the interfering diagrams are of similar size, rare decay modes which are not dominated by Cabibbo-favored tree-level transitions are good candidates to search for such effects. This includes in particular decays of *B* mesons to two light hadrons.

The direct *CP* asymmetry in the $b \to s\bar{s}s$ transition decay $B^+ \to \phi K^+$ is expected to be of the order of a few percent [9]. It was first measured at a hadron

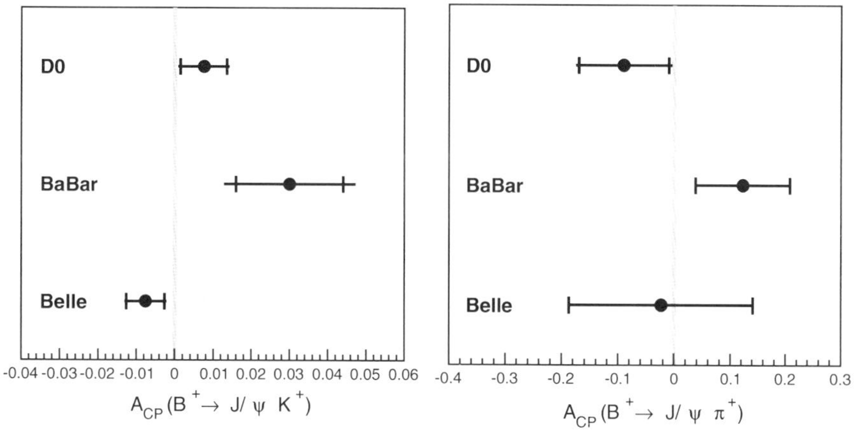

Fig. 8.2 Comparison of $A_{CP}(B^+ \to J/\psi K^+)$ (*left*) and $A_{CP}(B^+ \to J/\psi \pi^+)$ measurements (*right*) by D0 [4] with results from the B-factory experiments [5–8]

collider by CDF in a data sample corresponding to an integrated luminosity of 180 pb^{-1} [10].

The number of B^+ and B^- signal decays is determined in a fit to the B meson candidate invariant mass, the ϕ candidate invariant mass, the ϕ meson helicity angle, and the ionization energy loss measurement in the drift chamber. Background from random combinations of tracks and from several physics processes, like non-resonant $B^+ \to K^+K^-K^+$, are taken into account. The fit result is illustrated in the left plot of Fig. 8.3.

The measured asymmetry is $A_{CP}(B^+ \to \phi K^+) = -0.07 \pm 0.17\,(\text{stat})^{+0.03}_{-0.02}$ (syst) [10]. The systematic uncertainty includes the effect of different K^+ and K^- reconstruction efficiencies, but is dominated by the uncertainties arising from the fit model. The result agrees well with other measurements as shown in the right part of Fig. 8.3. So far the experimental precision is insufficient to figure out whether there is direct *CP* violation at the level of a few percent in this decay mode.

While several charmless decays of B^+ and B^0 mesons have been studied at the B-factories, the Tevatron extended the scope to B_s^0 mesons and Λ_b^0 baryons. One of the main challenges in the analysis of these rare decays is the statistical separation of the various two-body b-hadron modes. The method employed by CDF is based on a fit of three kinematic variables and a dE/dx measurement and is described in more detail in Sect. 9.1. In this Section we focus on the A_{CP} measurements of the charmless hadronic B decays.

The mode with the highest signal yield is $B^0 \to K^+\pi^-$. Its *CP* asymmetry was measured by CDF with a data sample of 180 pb^{-1} [13] and in an updated analysis using 1 fb^{-1} [14]. The increased data size allowed to also measure A_{CP} of the modes $B_s^0 \to K^-\pi^+$, $\Lambda_b^0 \to p\pi^-$, and $\Lambda_b^0 \to pK^-$ for the first time.

A projection of the fit from which the signal yields are extracted is shown in the left part of Fig. 8.4. To calculate the asymmetry from the yields they are corrected for the different trigger and reconstruction efficiency of the *CP* conjugated final states. For

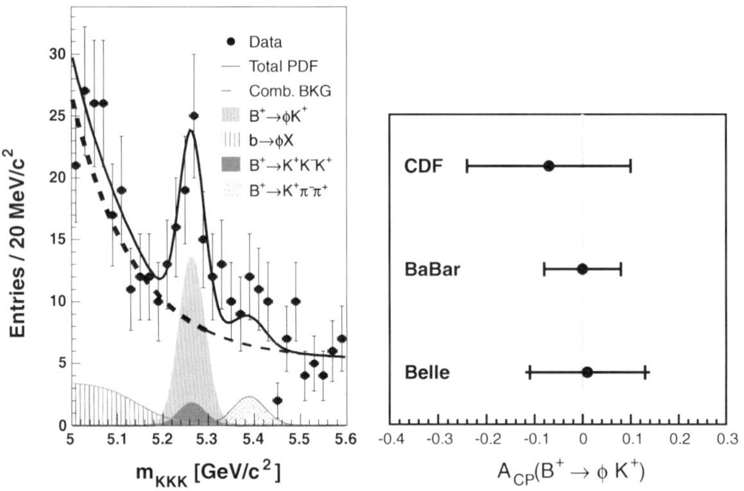

Fig. 8.3 Projection of the fit to the B meson candidate invariant mass [10] (*left*) and comparison of $A_{CP}(B^+ \to \phi K^+)$ results by CDF [10], BaBar [11], and Belle [12] (*right*)

the $K^+\pi^-$ final state the correction is measured in a high-statistics $D^0 \to K^-\pi^+$ sample where *CP*-violating effects are expected to be negligible and small differences in kinematics with respect to the signal mode are accounted for by simulation. The asymmetry between p and $\bar{p}$ efficiencies is measured in a similar way with $\Lambda \to p\pi^-$ decays.

Evidence for direct *CP* violation in $B^0 \to K^+\pi^-$ at the level of 3.5σ is obtained with an asymmetry of $A_{CP} = -0.086 \pm 0.023$ (stat)± 0.009 (syst) [14]. The results of $A_{CP}(B_s^0 \to K^-\pi^+) = +0.39 \pm 0.15$ (stat)± 0.08 (syst), and $A_{CP}(\Lambda_b^0 \to pK^-) = +0.37 \pm 0.17$ (stat) $\pm$ 0.03 (syst) differ from zero by 2.3 and 2.1σ, respectively. For $\Lambda_b^0 \to p\pi^-$ decays the asymmetry of $A_{CP} = +0.03 \pm 0.17$ (stat) $\pm$ 0.05 (syst) is well consistent with zero. The dominant systematic uncertainties of the A_{CP} results come from the description of the dE/dx measurement, the combinatorial background model, and b-hadron masses.

The $A_{CP}(B^0 \to K^+\pi^-)$ result confirms the large negative value measured at the B-factories as shown in Fig. 8.4. In Ref. [15] it was pointed out that this decay mode is related to $B_s^0 \to K^-\pi^+$ via $SU(3)$ flavor symmetry and a prediction of

$$R = \frac{A_{CP}(B^0 \to K^+\pi^-)}{A_{CP}(B_s^0 \to K^-\pi^+)} \frac{\mathscr{B}(B_s^0 \to K^-\pi^+)}{\mathscr{B}(B^0 \to K^+\pi^-)} \frac{\tau(B^0)}{\tau(B_s^0)} = 1 \tag{8.1}$$

is derived. The large central value of $A_{CP}(B_s^0 \to K^-\pi^+)$ is consistent with this prediction. CDF measures $R = 0.85 \pm 0.42$ (stat)± 0.13 (syst) [14] under the assumption of equal B_s^0 and B^0 lifetimes.

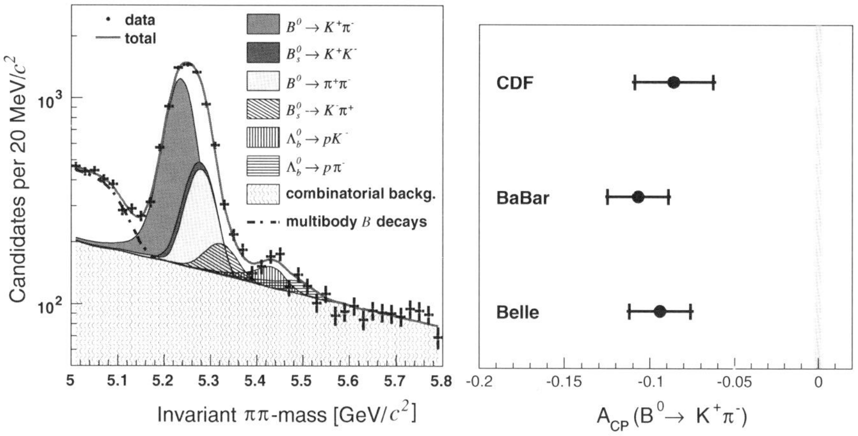

Fig. 8.4 Projection of the fit to the B meson candidate invariant mass assuming the pion mass hypothesis for both daughter particles [14] (*left*) and comparison of $A_{CP}(B^0 \to K^+\pi^-)$ results by CDF [14], BaBar [16], and Belle [1] (*right*)

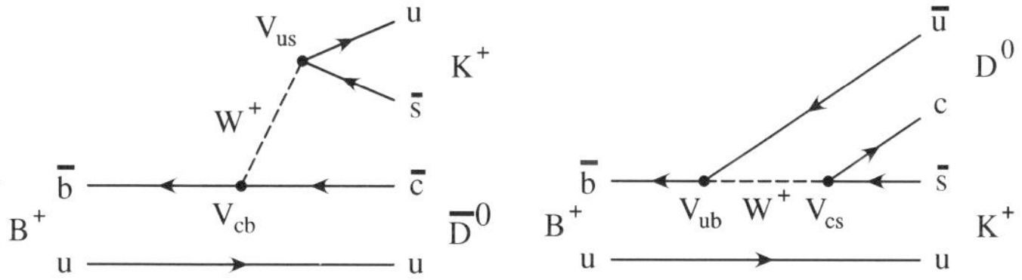

Fig. 8.5 Tree-level Feynman diagrams of the decay $B^+ \to DK^+$

8.3 $B \to DK$

Two tree-level diagrams contribute to the $B^+ \to DK^+$ decay. As one can see in Fig. 8.5, both are Cabibbo-suppressed of order λ^3, but the second process with the $\bar{b} \to \bar{u}c\bar{s}$ transition is in addition color-suppressed. This leads to an absolute ratio of amplitudes between color-suppressed and color-allowed decays,

$$r_B := \left| \frac{A(B^+ \to D^0 K^+)}{A(B^+ \to \bar{D}^0 K^+)} \right|, \tag{8.2}$$

of the order 0.1. The relative weak phase between the diagrams is $\arg(V_{cs} V_{ub}^* / V_{us} V_{cb}^*)$. As one can see from the Wolfenstein parametrization of the CKM matrix in Eq. (2.8) this phase is approximately equal to $\arg(-V_{ud} V_{ub}^* / V_{cd} V_{cb}^*) = \gamma$. The phase becomes accessible experimentally via a direct *CP* violation measurement if both diagrams interfere. Since the first process has a $\bar{D}^0$ in the final state and the second process a D^0, this is only possible if $\bar{D}^0$ and D^0 decay to the same final state.

One way to achieve this is to look for decays to *CP* eigenstates, like K^+K^- or $\pi^+\pi^-$. This method was proposed by Gronau, London and Wyler and is therefore called GLW method [17, 18]. The second method, proposed by Atwood, Dunietz and Soni and abbreviated ADS [19], exploits doubly Cabibbo-suppressed D^0 decays. Here the interference happens between the $B^+ \to \bar{D}^0 K^+$ decay followed by the doubly Cabibbo-suppressed $\bar{D}^0 \to K^-\pi^+$ decay and the color-suppressed $B^+ \to D^0 K^+$ transition with a Cabibbo-favored $D^0 \to K^-\pi^+$ decay. Finally, Giri, Grossman, Soffer, and Zupan proposed to measure the angle γ in $B^+ \to DK^+$ decays with three-body D meson final states, like $K^0_S\pi^+\pi^-$. The GGSZ method [20] requires a Dalitz plot analysis to measure the interference effects between Cabibbo-allowed, doubly Cabibbo-suppressed, and *CP*-eigenstate decay amplitudes. All methods assume that *CP* violation and mixing effects in the D^0 system are negligible. CDF contributed to the determination of the angle γ using the GLW and the ADS method.

In the GLW method, where the decay of the D meson to a *CP* eigenstate, $f_{CP\pm}$, is considered, the amplitudes A_1 and A_2 in Eq. (3.55) are given by

$$A_1 = A(B^+ \to \bar{D}^0 K^+)A(\bar{D}^0 \to f_{CP\pm}), \text{ and} \tag{8.3}$$

$$A_2 = A(B^+ \to D^0 K^+)A(D^0 \to f_{CP\pm}), \tag{8.4}$$

respectively. For a *CP*-even final state the amplitude $A(\bar{D}^0 \to f_{CP+})$ is equal to the amplitude of the *CP*-conjugated decay $D^0 \to f_{CP+}$ and it cancels in the decay rate asymmetry. With Eq. (8.2) follows that $|A_2| = r_B|A_1|$ and the rate asymmetry in Eq. (3.55) becomes

$$A_{CP+} = \frac{2r_b \sin\delta_B \sin\gamma}{1 + r_B^2 + 2r_B \cos\delta_B \cos\gamma}, \tag{8.5}$$

where δ_B is the relative strong phase between the two B meson decay amplitudes. Information on r_B and the phases can be obtained from a measurement of the average rate of decays with a *CP*-even final state of the D meson relative to the case with a Cabibbo-favored D decay

$$R_{CP+} = 2\frac{\Gamma(B^- \to D_{CP+}K^-) + \Gamma(B^+ \to D_{CP+}K^+)}{\Gamma(B^- \to D^0K^-) + \Gamma(B^+ \to \bar{D}^0K^+)}, \tag{8.6}$$

where D_{CP+} refers to the *CP*-even D meson eigenstate, $|D_{CP+}\rangle = \frac{1}{\sqrt{2}}(|D^0\rangle + |\bar{D}^0\rangle)$, and $\Gamma(B^+ \to D_{CP+}K^+) = \Gamma(B^+ \to D(\to f_{CP+})K^+)/\mathcal{B}(D^0 \to f_{CP+})$. With $\Gamma(B^+ \to \bar{D}^0K^+) \propto |A_1|^2$, $\Gamma(B^+ \to D_{CP+}K^+) \propto \frac{1}{2}|A_f|^2$, and $|A_f|^2$ and $|\bar{A}_{\bar{f}}|^2$ from Eqs. (3.53) and (3.54) this leads to

$$R_{CP+} = \frac{|A_f|^2 + |\bar{A}_{\bar{f}}|^2}{2|A_1|^2} = 1 + r_B^2 + 2r_B \cos\delta_B \cos\gamma \tag{8.7}$$

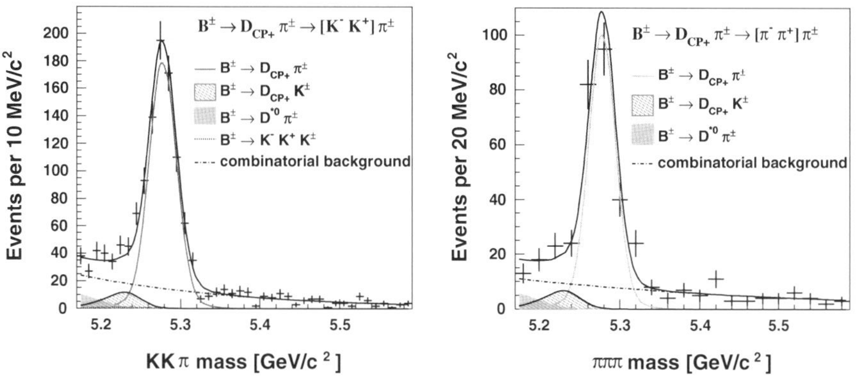

Fig. 8.6 Invariant mass distributions of $B^+ \to D\pi^+$ candidates with fit projections for the D meson decay modes K^+K^- and $\pi^+\pi^-$ [21]

and

$$A_{CP+} = 2r_b \sin\delta_B \sin\gamma / R_{CP+}. \tag{8.8}$$

CDF has measured A_{CP+} and R_{CP+} in a data sample of 1 fb^{-1} [21]. The D meson from the $B^+ \to DK^+$ decay is reconstructed in the CP-even final states K^+K^- and $\pi^+\pi^-$ and in the flavor-specific mode $K^+\pi^-$. For the R_{CP+} measurement the approximation $R_{CP+} = R_+/R$ with

$$R_+ = \frac{\mathscr{B}(B^- \to D_{CP+}K^-) + \mathscr{B}(B^+ \to D_{CP+}K^+)}{\mathscr{B}(B^- \to D_{CP+}\pi^-) + \mathscr{B}(B^+ \to D_{CP+}\pi^+)} \quad \text{and} \tag{8.9}$$

$$R = \frac{\mathscr{B}(B^- \to D^0K^-) + \mathscr{B}(B^+ \to \bar{D}^0K^+)}{\mathscr{B}(B^- \to D^0\pi^-) + \mathscr{B}(B^+ \to \bar{D}^0\pi^+)} \tag{8.10}$$

is used which relies on the well-motived assumption of no CP violation in the Cabibbo-favored $B^+ \to D\pi^+$ decay. The normalization to this decay in the double ratio reduces the systematic uncertainty related to the relative efficiency and branching ratio of the D decays to CP-even and flavor-specific final states.

The selection of $B^+ \to DK^+$ candidates is optimized to achieve the best resolution on A_{CP+}. One of the main challenges of the analysis is to separate the suppressed $B^+ \to DK^+$ decays from the Cabibbo-favored $B^+ \to D\pi^+$ decays. This is illustrated in Fig. 8.6, where the pion mass is assigned to the track from the B decay. A clear $B^+ \to D\pi^+$ peak is visible at the B^+ mass. The $B^+ \to DK^+$ events show up as a smaller peak shifted to lower mass values because of the pion mass assignment to the kaon.

The signal yields are determined in a fit to kinematic and particle identification variables. The fit method corresponds to the one used for the charmless b hadron decays discussed in the previous section and is described in Sect. 9.1. Since the radiative tail of the $B^+ \to D\pi^+$ signal from final state photon emissions overlaps

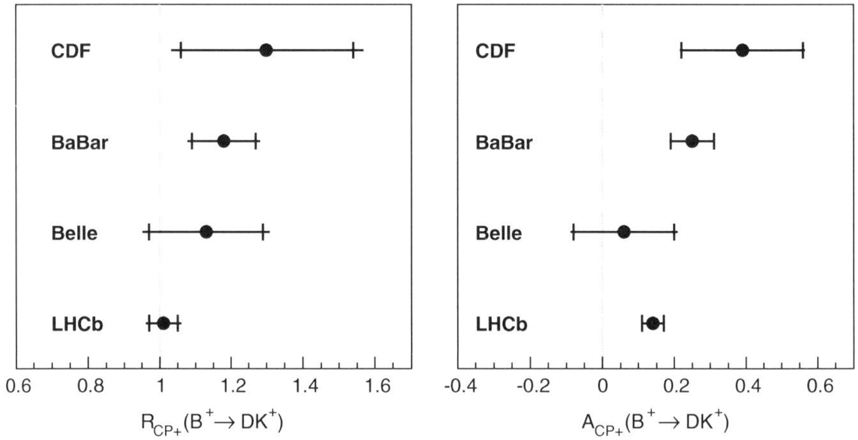

Fig. 8.7 R_{CP+} (*left*) and A_{CP+} values (*right*) measured in $B^+ \to DK^+$ decays at CDF [21], BaBar [22], Belle [23], and LHCb [24]

with the invariant mass peak of the $B^+ \to DK^+$ signal it has to be included in the mass model. Simulation is used to derive a detailed model which is verified on data with a high-statistics control sample. Besides the two signal components, combinatorial background and physics background, like $B^+ \to \bar{D}^{*0}\pi^+$ and non resonant $B^+ \to K^+K^-K^+$, are considered. A correction for the difference in K^+ and K^- reconstruction efficiency is obtained from simulation and verified on data.

The results are $R_{CP+} = 1.30 \pm 0.24$ (stat) $\pm$ 0.12 (syst) and $A_{CP+} = 0.39 \pm 0.17$ (stat) $\pm$ 0.04 (syst) [21]. The dominant systematic uncertainties come from the dE/dx model and the kinematic distributions of combinatorial background. The results have been the first such measurements at a hadron collider and agree with measurements at the B-factories and the new LHCb result as one can see in Fig. 8.7.

Since the GLW method exploits an interference effect between a color-allowed and a color-suppressed decay amplitude, the observable effect is limited by the amplitude ratio. As one can see in Eq. (8.5) the maximal A_{CP+} value is of the order of $2r_B$. The basic idea of the ADS method is to compensate this effect by taking a doubly Cabibbo-suppressed D decay for the color-allowed case which then interferes with the Cabibbo-allowed D decay of the color-suppressed B decay process. A drawback of the method is that the two D meson decays are not related by *CP* symmetry any more and two further parameters have to be introduced, the absolute ratio and the relative strong phase between $D^0 \to f$ and $\bar{D}^0 \to f$ decay amplitudes,

$$r_D = \left|\frac{A(\bar{D}^0 \to f)}{A(D^0 \to f)}\right|, \quad \text{and } \delta_D = \arg\left(\frac{A(\bar{D}^0 \to f)}{A(D^0 \to f)}\right), \tag{8.11}$$

where f is a flavor-specific final state of a Cabibbo-favored D^0 decay. While r_D is precisely measured for the $K^-\pi+$ final state, the phase δ_D is only known with large uncertainties.

The relative rates and asymmetry of $B^+ \to DK^+$ decays with a suppressed D decay ($D^0 \to \bar{f}$ or $\bar{D}^0 \to f$) are then related to the parameters of interest via

$$\begin{aligned} R &= \frac{\Gamma(B^- \to D(\to \bar{f})K^-) + \Gamma(B^+ \to D(\to f)K^+)}{\Gamma(B^- \to D(\to f)K^-) + \Gamma(B^+ \to D(\to \bar{f})K^+)} \\ &= r_B^2 + r_D^2 + 2r_B r_D \cos(\delta_B + \delta_D)\cos\gamma \end{aligned} \tag{8.12}$$

$$\begin{aligned} R^+ &= \frac{\Gamma(B^+ \to D(\to f)K^+)}{\Gamma(B^+ \to D(\to \bar{f})K^+)} \\ &= r_B^2 + r_D^2 + 2r_B r_D \cos(\delta_B + \delta_D + \gamma) \end{aligned} \tag{8.13}$$

$$\begin{aligned} R^- &= \frac{\Gamma(B^- \to D(\to \bar{f})K^-)}{\Gamma(B^- \to D(\to f)K^-)} \\ &= r_B^2 + r_D^2 + 2r_B r_D \cos(\delta_B + \delta_D - \gamma) \end{aligned} \tag{8.14}$$

$$\begin{aligned} A &= \frac{\Gamma(B^- \to D(\to \bar{f})K^-) - \Gamma(B^+ \to D(\to f)K^+)}{\Gamma(B^- \to D(\to \bar{f})K^-) + \Gamma(B^+ \to D(\to f)K^+)} \\ &= 2r_B r_D \sin(\delta_B + \delta_D)\sin\gamma / R. \end{aligned} \tag{8.15}$$

All four parameters were measured by CDF using 7 fb^{-1} of data [25]. Suppressed $B^+ \to DK^+$ events are selected by requiring the D meson to decay to $K^-\pi^+$. If the D meson candidate is also consistent with the Cabibbo-favored $K^+\pi^-$ hypothesis it is rejected from the sample of suppressed decays. Favored $B^+ \to DK^+$ decays with $D \to K^+\pi^-$ are selected correspondingly. The symmetry of the selection for both modes leads to the cancellation of systematic uncertainties in the ratio. Background contributions from non-resonant three-body B^+ decays and from B decays where a B and a D daughter particle are interchanged are reduced by requirements on the distance between B and D decay vertex and the invariant mass of the B daughter track and the oppositely charged D daughter track. The selection requirements on several other variables are optimized on $S/(1.5 + \sqrt{B})$ [26], where S and B are the expected number of signal and background events in the sample of suppressed decays, respectively. This figure of merit maximizes the sensitivity for a signal with a significance of 3σ.

The signal yields are determined in a fit to the invariant B^+ candidate mass, where the pion mass hypothesis is used for the B^+ daughter track, like in the GLW analysis, and the ionization energy loss measurement in the drift chamber. Backgrounds arise from random combinations of tracks and from partially reconstructed or non-resonant B decays. Fit projections are shown in Fig. 8.8. Correction factors for different reconstruction efficiencies of K^+, K^-, π^+, and π^- are measured in data.

A signal of suppressed $B^+ \to DK^+$ decays with $D \to K^-\pi^+$ is observed with a significance of 3.2σ. The measured relative rates and asymmetry are $R = [22.0 \pm 8.6\,(\text{stat}) \pm 2.6\,(\text{syst})] \times 10^{-3}$, $R^+ = [42.6 \pm 13.7\,(\text{stat}) \pm 2.8\,(\text{syst})] \times 10^{-3}$, $R^- = [3.8 \pm 10.3\,(\text{stat}) \pm 2.7\,(\text{syst})] \times 10^{-3}$, and $A = -0.82 \pm 0.44\,(\text{stat}) \pm 0.09\,(\text{syst})$

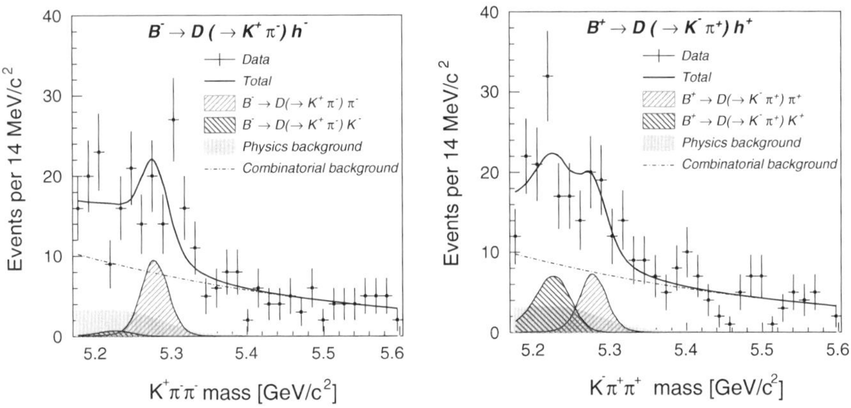

Fig. 8.8 Invariant mass of $B^- \to D(\to K^+\pi^-)\pi^-$ (*left*) and $B^+ \to D(\to K^-\pi^+)\pi^+$ candidates (*right*) with fit projection [25]

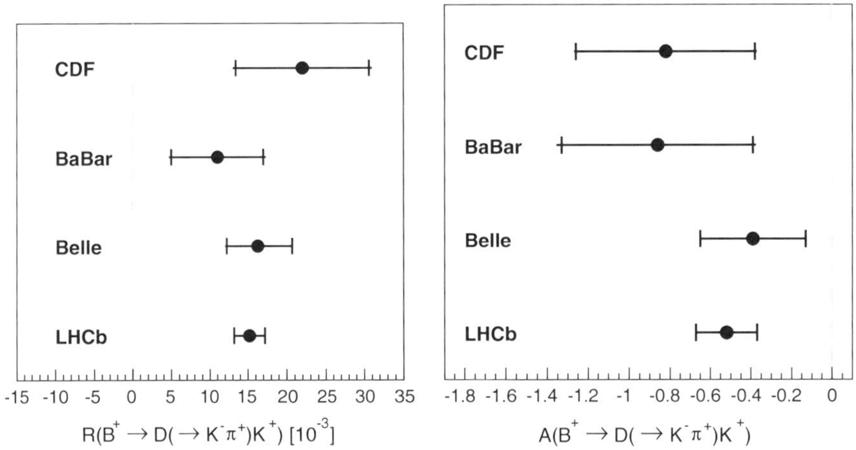

Fig. 8.9 *R* (*left*) and *A* values (*right*) measured in $B^+ \to DK^+$ decays with $D \to K^-\pi^+$ at CDF [25], BaBar [27], Belle [28], and LHCb [24]

[25]. Results for the $B^+ \to D\pi^+$ decay are reported as well, but no indication of *CP* violation is found in this mode.

Figure 8.9 shows a comparison of the *R* and *A* results with other measurements. The situation is similar to the GLW results. The measurements agree and the lead in precision is now taken by LHCb. The larger central value of *A* compared to A_{CP+} determined in the GLW analyses supports the assumption of a stronger *CP* violating effect. But because of the lower signal yield due to the doubly Cabibbo-suppressed *D* decay the uncertainty is larger as well so that the *CP* violating effect is less significant.

8.4 Charm Meson Decays

The singly Cabibbo-suppressed decay modes $D^0 \to K^+K^-$ and $D^0 \to \pi^+\pi^-$ have been suggested to search for *CP* violation in the charm sector [29–33]. Both modes could show *CP* violation in the decay via the interference of the tree-level and penguin diagrams. But the penguin contribution is expected to be tiny in the standard model. Since the final state is a *CP*-even eigenstate accessible by D^0 and $\bar{D}^0$ mesons, mixing-induced *CP* violation can be present, too. But also here the effect is expected to be small in the standard model. *CP* asymmetries of the order of a percent or higher would usually be considered a hint for new physics contributions. On the other hand theoretical predictions for charm mesons have larger uncertainties than calculations for kaons and *B* mesons because the charm quark mass is too heavy to be ignored, but it provides a lower factorization scale than the *b* quark mass.

A time integrated measurement of *CP* asymmetries of D^0 decays to h^+h^-, where h stands for π or K, was performed by CDF on a data set of 123 pb^{-1} [34] and in an updated analysis on 5.9 fb^{-1} [35]. The new analysis does not only profit from the increased data size, but also uses refined techniques which are described below to reduce the systematic uncertainties.

As in the case of the charm mixing analysis presented in Sect. 6.3, the flavor of the *D* meson at production is tagged by the charge of the pion from a $D^{*+} \to D^0\pi^+$ decay, called slow pion. A critical aspect of this analysis is to control the asymmetry in the reconstruction efficiency between positive and negative pions from the D^* decay. The asymmetry is caused by different interaction cross sections for low momentum pions and the tilted geometry of the cells in the drift chamber to compensate the Lorentz angle. A data-driven method is employed to measure the detection asymmetries. If all asymmetries are small so that higher order terms can be neglected, the observed asymmetry for $D \to h^+h^-$ decays, $A(h^+h^-)$, is given by the sum of the *CP* asymmetry, $A_{CP}(D \to h^+h^-)$, and the instrumental asymmetry of the slow pions, $\delta(\pi_s)$:

$$A(h^+h^-) = A_{CP}(D \to h^+h^-) + \delta(\pi_s). \tag{8.16}$$

To determine $\delta(\pi_s)$, Cabibbo-favored $D^0 \to K^-\pi^+$ and $\bar{D}^0 \to K^+\pi^-$ decays are used. If such decays are tagged by a slow pion from a D^* decay and the slow pion has the same kinematic distributions as in the tagged $D \to h^+h^-$ decays, the observed asymmetry, $A_{\text{tagged}}(K^-\pi^+)$, is affected by the same slow pion asymmetry, $\delta(\pi_s)$:

$$A_{\text{tagged}}(K^-\pi^+) = A_{CP}(D \to K^-\pi^+) + \delta(\pi_s) + \delta(K\pi). \tag{8.17}$$

In addition the instrumental asymmetry between the reconstruction efficiencies of $K^-\pi^+$ and $K^+\pi^-$, $\delta(K\pi)$, enters. The contributions of $A_{CP}(D \to K^-\pi^+)$ and $\delta(K\pi)$ can be measured in untagged $D^0 \to K^-\pi^+$ decays, again assuming equal D^0 kinematics in the tagged and untagged samples:

$$A_{\text{untagged}}(K^-\pi^+) = A_{CP}(D \to K^-\pi^+) + \delta(K\pi). \tag{8.18}$$

Combining Eqs. (8.16), (8.17), and (8.18) leads to

$$A_{CP}(D \to h^+h^-) = A(h^+h^-) - A_{\text{tagged}}(K^-\pi^+) + A_{\text{untagged}}(K^-\pi^+). \quad (8.19)$$

The above consideration relies on the fact that there is no production asymmetry which should hold for charm hadrons produced in strong interactions of $p\bar{p}$ collisions. It is also assumed that the efficiencies for the slow pion and the $K\pi$ pair factorize. Systematic uncertainties for both assumptions are assigned.

The method requires equal kinematic distributions of the slow pion in the h^+h^- and $K\pi$ mode and of the D^0 in the tagged and untagged $D^0 \to K^-\pi^+$ sample. To ensure this several kinematic variables are checked and the events are reweighted in the distributions where differences are observed. These are the transverse momentum and the pseudorapidity of the slow pion and the D^0.

Signal event yields in the tagged samples are determined in a fit to the reweighted D^* candidate invariant mass distribution. The D^* candidate invariant mass is calculated with the D^0 candidate invariant mass fixed to the known D^0 mass. This results in the same mass resolution as obtained from the often used mass difference between D^* and D^0 candidate, and has the additional advantage that the signal shape becomes independent of the D^0 decay mode. Random combinations of tracks and multi-body D decays are considered as backgrounds. Fit results are shown in Fig. 8.10. The yields of untagged $D^0 \to K^-\pi^+$ and $\bar{D}^0 \to K^+\pi^-$ decays are extracted from a fit to the $K\pi$ invariant mass distribution. The fit includes an additional component for D decays with wrongly assigned kaon and pion hypotheses and is performed on two independent samples to avoid correlations.

The measured *CP* asymmetries corrected for detector effects are $A_{CP}(D \to K^+K^-) = [-0.24 \pm 0.22\,(\text{stat}) \pm 0.09\,(\text{syst})]\,\%$ and $A_{CP}(D \to \pi^+\pi^-) = [0.22 \pm 0.24\,(\text{stat}) \pm 0.11\,(\text{syst})]\,\%$ [35]. The dominant systematic uncertainty comes from the mass model and the difference in the mass model between D^0 and $\bar{D}^0$. Both results are consistent with the previous CDF measurement and the B-factory results as shown in Fig. 8.11. No indication of new physics is found.

As discussed before, the measured A_{CP} values can contain contributions from direct and indirect *CP* violation. While the former is time-independent, the latter depends on the D^0 decay time. Because of the slow D^0 oscillation the time dependence can be approximated well by the linear term so that the observed A_{CP} value becomes

$$A_{CP}(D \to h^+h^-)(t) = A_{CP}^{\text{dir}}(D \to h^+h^-) + \frac{t}{\tau} A_{CP}^{\text{ind}}(D \to h^+h^-). \quad (8.20)$$

If the direct *CP* violation is assumed to be negligible, the contribution from indirect *CP* violation can be calculated from the measured A_{CP} value and the average decay time of observed D^0 mesons, $\langle t \rangle$, which depends on the detector acceptance. The $\langle t \rangle$ values are determined in a fit to sideband-subtracted decay time distributions to account for D^0 mesons from B decays. The high values of about 2.5 D^0 lifetimes are a consequence of the two-track trigger requirements and imply an increased

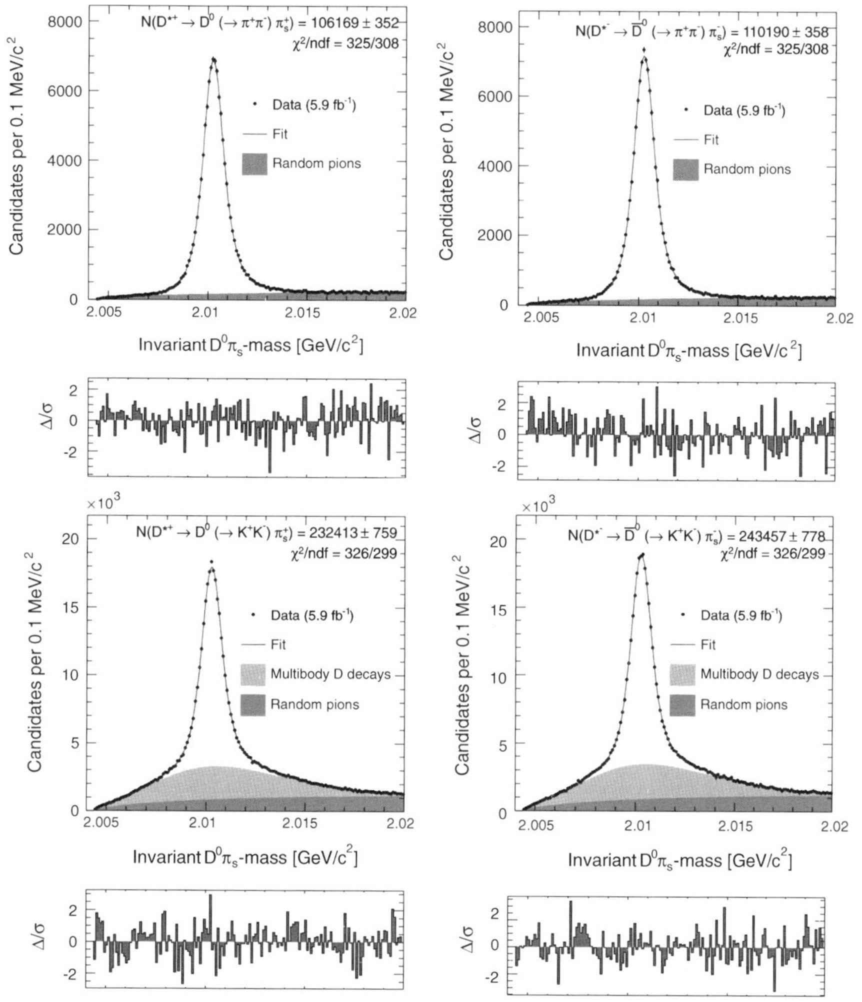

Fig. 8.10 Result of the fit to the D^* candidate invariant mass for $D^0 \to \pi^+\pi^-$ (*top row*) and $D^0 \to K^+K^-$ (*bottom row*) [35]. The left plots show the D^0 and the right plots the $\bar{D}^0$ candidates

sensitivity to mixing-induced *CP* violation compared to a uniform trigger acceptance as at the B-factories. Assuming $A_{CP}^{\text{dir}}(D \to h^+h^-) = 0$, the results $A_{CP}^{\text{ind}}(D \to K^+K^-) = [-0.08 \pm 0.08\,(\text{stat}) \pm 0.03\,(\text{syst})]\,\%$ and $A_{CP}^{\text{ind}}(D \to \pi^+\pi^-) = [0.09 \pm 0.10\,(\text{stat}) \pm 0.05\,(\text{syst})]\,\%$ [35] are obtained. As indirect *CP* violation is independent of the decay mode both measurements can be combined to $A_{CP}^{\text{ind}}(D \to h^+h^-) = [-0.01 \pm 0.06\,(\text{stat}) \pm 0.04\,(\text{syst})]\,\%$.

Finally, the difference between the A_{CP} values of $\Delta A_{CP}(D \to h^+h^-) = A_{CP}(D \to K^+K^-) - A_{CP}(D \to \pi^+\pi^-) = [-0.46 \pm 0.31\,(\text{stat}) \pm 0.13\,(\text{syst})]\,\%$ [35] is quoted. Since the $\langle t \rangle$ values for the K^+K^- and $\pi^+\pi^-$ modes are similar the

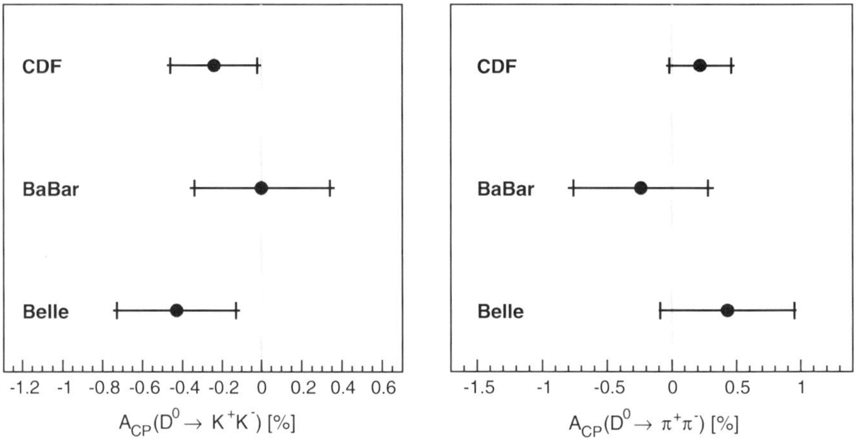

Fig. 8.11 $A_{CP}(D \to K^+K^-)$ (*left*) and $A_{CP}(D \to \pi^+\pi^-)$ measurements (*right*) CDF [35], BaBar [36], and Belle [37]

contributions from indirect *CP* violation approximately cancel in ΔA_{CP}. The central value differs from zero by about half a percent, but is still consistent with it within the uncertainties. A significant deviation of ΔA_{CP} from zero was recently reported by LHCb [2]. Whether this is caused by new physics has to be clarified in further measurements and improved theoretical calculations.

References

1. S. Lin et al., Nature **452**, 332 (2008)
2. R. Aaij et al., Phys. Rev. Lett. **108**, 111602 (2012)
3. W.S. Hou, *M* (A. Soddu, Nagashima, 2006)
4. V.M. Abazov et al., Phys. Rev. Lett. **100**, 211802 (2008)
5. B. Aubert et al., Phys. Rev. Lett. **94**, 141801 (2005)
6. K. Sakai et al., Phys. Rev. **D82**, 091104 (2010)
7. B. Aubert et al., Phys. Rev. Lett. **92**, 241802 (2004)
8. K. Abe et al., Phys. Rev. **D67**, 032003 (2003)
9. M. Beneke, M. Neubert, Nucl. Phys. **B675**, 333 (2003)
10. D.E. Acosta et al., Phys. Rev. Lett. **95**, 031801 (2005)
11. B. Aubert et al., Phys. Rev. **D74**, 032003 (2006)
12. K. Chen et al., Phys. Rev. Lett. **91**, 201801 (2003)
13. A. Abulencia et al., Phys. Rev. Lett. **97**, 211802 (2006)
14. T. Aaltonen et al., Phys. Rev. Lett. **106**, 181802 (2011)
15. H.J. Lipkin, Phys. Lett. B**621**, 126 (2005)
16. B. Aubert et al., Phys. Rev. Lett. **99**, 021603 (2007)
17. M. Gronau, D. London, Phys. Lett. B**253**, 483 (1991)
18. M. Gronau, D. Wyler, Phys. Lett. B**265**, 172 (1991)
19. D. Atwood, I. Dunietz, A. Soni, Phys. Rev. Lett. **78**, 3257 (1997)
20. A. Giri, Y. Grossman, A. Soffer, J. Zupan, Phys. Rev. **D68**, 054018 (2003)

21. T. Aaltonen et al., Phys. Rev. **D81**, 031105 (2010)
22. P. del Amo Sanchez et al., Phys. Rev. **D82**, 072004 (2010)
23. K. Abe et al., Phys. Rev. **D73**, 051106 (2006)
24. R. Aaij et al., Phys. Lett. B**712**, 203 (2012)
25. T. Aaltonen et al., Phys. Rev. **D84**, 091504 (2011)
26. G. Punzi, eConf C030908, MODT002 (2003)
27. P. del Amo Sanchez et al., Phys. Rev. **D82**, 072006 (2010)
28. Y. Horii et al., Phys. Rev. Lett. **106**, 231803 (2011)
29. D.s. Du, Eur. Phys. J. **C50**, 579 (2007)
30. Y. Grossman, A.L. Kagan, Y. Nir, Phys. Rev. **D75**, 036008 (2007)
31. I. Bigi, UND-HEP-09-BIG01 (2009)
32. S. Bianco, F. Fabbri, D. Benson, I. Bigi, Riv. Nuovo Cim. **26N7**, 1 (2003)
33. Z.z. Xing, Phys. Rev. **D55**, 196 (1997)
34. D.E. Acosta et al., Phys. Rev. Lett. **94**, 122001 (2005)
35. T. Aaltonen et al., Phys. Rev. **D85**, 012009 (2012)
36. B. Aubert et al., Phys. Rev. Lett. **100**, 061803 (2008)
37. M. Staric et al., Phys. Lett. **B670**, 190 (2008)

Chapter 9
Rare Decays

So far all experimental data is described rather well by the standard model. This means that any physics beyond the standard model can have only small effects. To search for such effects it is thus advantageous to look at processes that are suppressed or even forbidden in the standard model. The rare decay measurements and searches at the Tevatron are presented in this Chapter.

9.1 Charmless Hadronic B Decays

The dominant modes of weakly decaying b hadrons are $b \rightarrow c$ transitions. If the final states does not contain any open or hidden charm, it is suppressed in the standard model. *CP* asymmetry measurements for such decay mode were already discussed in Sect. 8.2. In this section, the experimental technique is explained and branching ratio measurements are presented.

The considered modes are decays of B^0, B_s^0, and Λ_b^0, denoted B in the following, to two charged hadrons, $h^+h'^-$, where $h^{(\prime)}$ stands for a pion, kaon, or proton. The two-body decays are selected by the Two-Track-Trigger of the CDF experiment. Background is suppressed by a requirement on the vertex fit quality and the isolation of the b hadron candidates. The isolation is defined as

$$I = \frac{p_T(B)}{p_T(B) + \sum_i p_{T,i}}, \tag{9.1}$$

where $p_T(B)$ is the transverse momentum of the b hadron candidates and the sum runs over all tracks, except for the B daughters, in a cone with radius 1 in $\eta\phi$ space around the B momentum direction. The hard fragmentation of b-quarks leads to the production of only few additional particles around the B meson and thus to a high isolation value.

For each B candidate, the invariant mass of the track pair, $m_{\pi\pi}$, is calculated under the assumption of the pion mass hypothesis for both daughter particles. As one can

T. Kuhr, *Flavor Physics at the Tevatron*, Springer Tracts in Modern Physics 249,
DOI: 10.1007/978-3-642-10300-1_9, © Springer-Verlag Berlin Heidelberg 2013

see in Fig. 9.1, the distributions of several $B \to h^+h'^-$ modes overlap. In addition combinatorial background from random combinations of tracks and physics background from partially reconstructed multibody B decays are present. To statistically separate the contributions of each mode, an unbinned likelihood fit in three kinematic and one particle identification variable is performed. Besides $m_{\pi\pi}$, the kinematic variables are the momentum imbalance, α, and the total momentum,

$$p_{tot} = p_1 + p_2 \tag{9.2}$$

where p_1 and p_2 are the momenta of the daughter particles. With p_1 being the lower and p_2 the higher of the two momenta, the momentum imbalance is defined as

$$\alpha = q_1 \frac{p_2 - p_1}{p_2}, \tag{9.3}$$

where q_1 is the charge of the lower momentum particle. The momenta of the daughter particles are then given by

$$p_1 = \frac{1 - |\alpha|}{2 - |\alpha|} p_{tot}, \quad p_2 = \frac{1}{2 - |\alpha|} p_{tot}. \tag{9.4}$$

Instead of the momentum imbalance, the momentum asymmetry, $\beta = -q_1(p_2 - p_1)/p_{tot}$, is used in some analyses. For a given decay of a mother particle with mass m_{12} to two daughter particles with masses m_1 and m_2, the three kinematic variables are correlated via

$$m_{12}^2 - m_{\pi\pi} = m_1^2 + m_2^2 - 2m_\pi^2 + 2\sqrt{p_1^2 + m_1^2}\sqrt{p_2^2 + m_2^2} - 2\sqrt{p_1^2 + m_\pi^2}\sqrt{p_2^2 + m_\pi^2}. \tag{9.5}$$

In the relativistic limit the dependency on p_{tot} vanishes:

$$\begin{aligned} m_{12}^2 - m_{\pi\pi} &= (1 - p_2/p_1)(m_1^2 - m_\pi^2) + (1 - p_1/p_2)(m_2^2 - m_\pi^2) \\ &= (2 - |\alpha|)(m_1^2 - m_\pi^2) + (1 - (|\alpha| - 1)^{-1})(m_2^2 - m_\pi^2). \end{aligned} \tag{9.6}$$

The relation between $m_{\pi\pi}$ and α is illustrated in the right plot of Fig. 9.1.

The PDFs for the kinematic distributions of the signal components are derived from simulation. Final state radiation is simulated by the PHOTOS package [3]. The mass resolution model was checked on high-statistics $D^0 \to K^-\pi^+$ and $D^0 \to \pi^+\pi^-$ samples.

Particle identification information is obtained from the ionization energy loss measurement in the drift chamber. The dE/dx response for pions and kaons is calibrated with $D^0 \to K^-\pi^+$ events where the D^0 flavor is tagged by a $D^{*+} \to D^0\pi^+$ decay. The calibration for protons is obtained from $\Lambda \to p\pi^-$ decays.

Branching ratios are measured relative to $B^0 \to K^+\pi^-$ which is the $B \to h^+h'^-$ mode with the highest signal yield and whose absolute branching ratio is measured

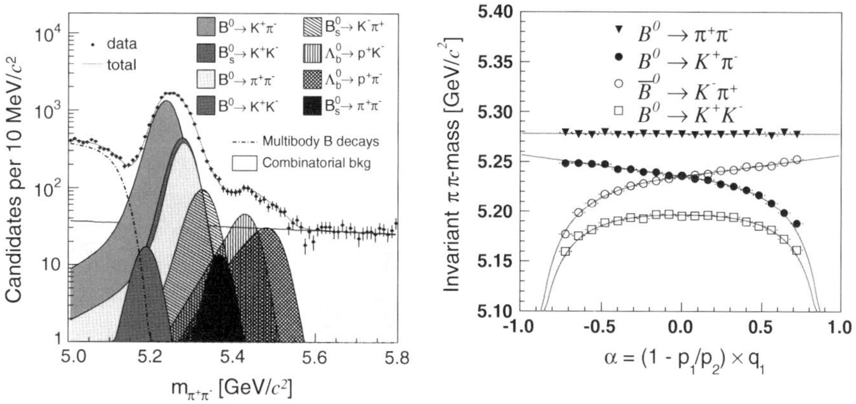

Fig. 9.1 Distribution of $m_{\pi\pi}$ with fit projection [1] (*left*) and dependency of $m_{\pi\pi}$ on the momentum imbalance α [2] (*right*)

at the B-factories. The relative selection efficiency between signal and normalization modes is taken from simulation, except for two factors. The relative efficiency of the isolation requirement is measured with $B_s^0 \to J/\psi\phi$ and $B^0 \to J/\psi K^{*0}$ decays. The difference in trigger efficiency between kaons and pions due to their different specific ionization is obtained from $D^0 \to K^+K^-$, $D^0 \to K^-\pi^+$, and $D^0 \to \pi^+\pi^-$ decays.

A main source of the systematic uncertainties in the relative branching ratio measurements is the dE/dx model and calibration. A further important contribution to the systematic error comes from the description of the background.

The charmless hadronic decays $B_s^0 \to K^+K^-$, $B_s^0 \to K^-\pi^+$, $\Lambda_b^0 \to pK^-$, and $\Lambda_b^0 \to p\pi^-$ can proceed via a tree-level $b \to u$ transition. Thus they are sensitive to the magnitude and phase of V_{ub}. Rare B^0 decays that are related to B_s^0 decay via $SU(3)$ flavor symmetry can be used to constrain hadronic effects.

The $B_s^0 \to K^+K^-$ decay was first observed by CDF in a data sample of 180 pb^{-1} [2]. The relative branching ratio measured with 1 fb^{-1} of data is $f_s/f_d\mathscr{B}(B_s^0 \to K^+K^-)/\mathscr{B}(B^0 \to K^+\pi^-) = 0.347 \pm 0.020\,(\text{stat}) \pm 0.021\,(\text{syst})$ [4], where f_s/f_d is the ratio of B_s^0 to B^0 production cross sections. The result agrees with theoretical predictions [5–9]. The uncertainties of the predictions which are larger than the experimental one do not allow to discriminate between the models.

A further CDF analysis of 1 fb^{-1} of data yielded the first observations of the decays $B_s^0 \to K^-\pi^+$, $\Lambda_b^0 \to pK^-$, and $\Lambda_b^0 \to p\pi^-$ [10]. The latter two decays were already searched for by CDF using a mass fit only on 193 pb^{-1} of data where no significant signals were observed [11]. The measured relative branching ratios are $f_s/f_d\mathscr{B}(B_s^0 \to K^-\pi^+)/\mathscr{B}(B^0 \to K^+\pi^-) = 0.071 \pm 0.010\,(\text{stat}) \pm 0.007\,(\text{syst})$, $f_\Lambda/f_d\mathscr{B}(\Lambda_b^0 \to pK^-)/\mathscr{B}(B^0 \to K^+\pi^-) = 0.066 \pm 0.009\,(\text{stat}) \pm 0.008\,(\text{syst})$, and $f_\Lambda/f_d\mathscr{B}(\Lambda_b^0 \to p\pi^-)/\mathscr{B}(B^0 \to K^+\pi^-)$

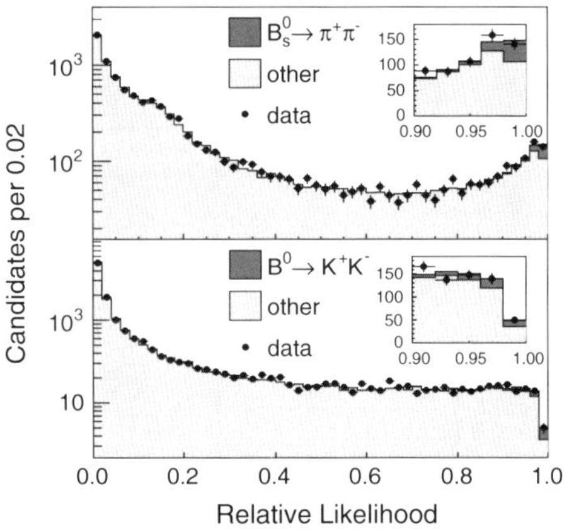

Fig. 9.2 Distribution of $\mathscr{L}_S/(\mathscr{L}_S + \mathscr{L}_{other})$, where $\mathscr{L}_S$ is the likelihood for the signal hypothesis and $\mathscr{L}_{other}$ is the likelihood for everything but the signal, for $B_s^0 \to \pi^+\pi^-$ (*top*) and $B^0 \to K^+K^-$ (*bottom*) [1]

$= 0.042 \pm 0.007$ (stat) $\pm$ 0.006 (syst) [10], where f_Λ/f_d is the ratio of Λ_b^0 to B^0 production rates.

Components for the decays $B_s^0 \to \pi^+\pi^-$ and $B^0 \to K^+K^-$ are included in the fit as well, but no significant signals are observed and limits on their branching ratio are quoted. First evidence for the decay $B_s^0 \to \pi^+\pi^-$ with a significance of 3.7σ and a 2.0σ signal of the $B^0 \to K^+K^-$ decay were reported by CDF in an analysis of 6 fb^{-1} [1]. The signals are visualized in Fig. 9.2 in a plot of the relative signal likelihood which contains the information from all four observables used in the fit. The results are $f_s/f_d\mathscr{B}(B_s^0 \to \pi^+\pi^-)/\mathscr{B}(B^0 \to K^+\pi^-) = 0.008\pm0.002$ (stat)$\pm$ 0.001 (syst) and $\mathscr{B}(B^0 \to K^+K^-)/\mathscr{B}(B^0 \to K^+\pi^-) = 0.012 \pm 0.005$ (stat) $\pm$ 0.005 (syst) [1].

The decays $B_s^0 \to \pi^+\pi^-$ and $B^0 \to K^+K^-$ are of particular interest because all quarks in the final state are different from the quarks in the initial state so that the decays can only proceed via annihilation diagrams. A penguin annihilation and a weak exchange diagram for the $B_s^0 \to \pi^+\pi^-$ decay are shown in Fig. 9.3. A comparison between $B_s^0 \to \pi^+\pi^-$ and $B^0 \to K^+K^-$ branching ratios may provide information about the relative contributions of both diagrams and improve the understanding of annihilation processes.

9.2 $B_{(s)}^0 \to \ell^+\ell^-$

Leptonic decays of B^0 and B_s^0 mesons are flavor changing neutral current (FCNC) processes. In the standard model such transitions are forbidden at tree level and require higher order box or penguin diagrams as shown in Fig. 9.4. The decay is also suppressed by the involved CKM matrix elements. For B^0 mesons the suppression is

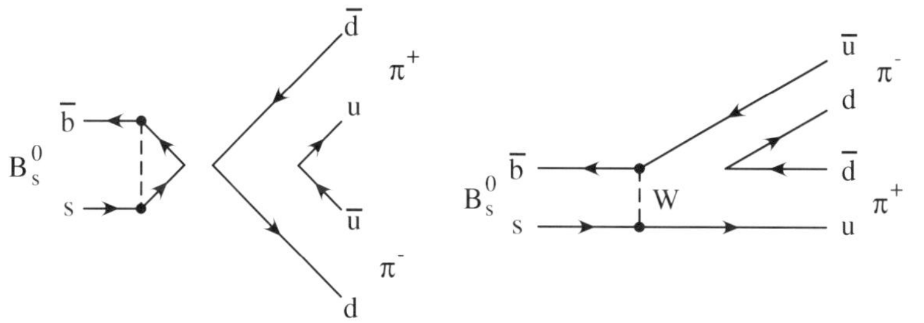

Fig. 9.3 Penguin annihilation (*left*) and a weak exchange diagrams (*right*) for the decay $B^0_s \to \pi^+\pi^-$

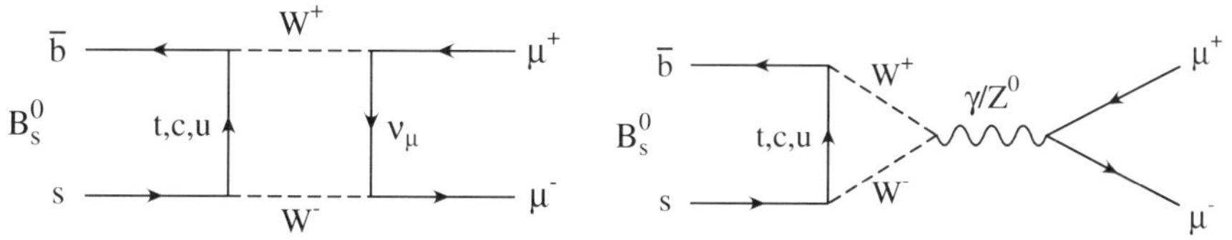

Fig. 9.4 Feynman diagrams of the $B^0_s \to \mu^+\mu^-$ decay

stronger compared to B^0_s decays by a factor of $|V_{td}/V_{ts}|^2 \approx 0.04$. A further reduction of the partial decay width comes from the helicity suppression because the spinless B meson decays two a left-handed lepton and a right-handed anti-lepton.

The branching ratios of the decays to two muons can be calculated precisely in the standard model and are $\mathcal{B}(B^0_s \to \mu^+\mu^-) = (3.2 \pm 0.2) \times 10^{-9}$ and $\mathcal{B}(B^0 \to \mu^+\mu^-) = (1.0 \pm 0.1) \times 10^{-10}$ [12, 13]. By taking into account the effect of the finite decay width difference in the B^0_s system, a 10 % higher value of $\mathcal{B}(B^0_s \to \mu^+\mu^-)$ is derived in Ref. [14]. The small value and rather precise prediction of the $B^0_s \to \mu^+\mu^-$ branching ratio make this decay an excellent probe for new physics contributions. Several models had predicted large enhancements of the branching ratio, sometimes even orders of magnitude above the standard model value. Among these model are supersymmetry with minimal flavor violation [15], or new flavor dynamics [16, 17], and minimal supergravity [18]. With the increasing precision of the measurements the parameter space of the new physics models has been tightly constrained [19] and some models could even be excluded.

The general strategy to search for $B^0_s \to \mu^+\mu^-$ and $B^0 \to \mu^+\mu^-$ decays is the following. Two oppositely charged muon track candidates are combined and fitted to a common vertex. Signal events are identified by a peak in the invariant dimuon mass, $m_{\mu\mu}$. Backgrounds that contribute to the selected sample of $B^0_{(s)} \to \mu^+\mu^-$ candidates are random combinations of two tracks, two muons from a sequential B hadron decay $b \to c\mu^- \to \mu^+\mu^- X$, and $B \to h^+h'^-$ decays where both hadrons are mis-identified as muons. The combinatorial background consists mainly of two muons from two different heavy hadron decays and has a smooth $m_{\mu\mu}$ distribution. Because of the partial reconstruction of the sequential b hadron decays they populate the invariant mass region below about 5 GeV/c^2. The $B \to h^+h'^-$ background peaks in the signal region, but is slightly shifted to lower masses because the pion and kaon daughter tracks are assigned the muon mass.

To separate signal from background events a selection on several variables is applied. Track pairs originating from soft QCD processes are reduced by requirements on the transverse momenta of the B or muon candidate. The long lifetime of B mesons is exploited in variables like the decay length or the impact parameter of the muons and their significances. Sequential decays can be suppressed by requiring a small pointing angle between the reconstructed momentum and flight direction of the B candidate. Random combinations of two muons from two different heavy hadron decays can be identified by a bad vertex fit quality. The hard fragmentation of b-quarks is exploited by the isolation variable defined in Eq. (9.1). Finally particle identification information helps to reduce the contribution from mis-identified hadrons. The selection criteria are optimized without looking at the distribution of data events in the invariant mass region where the signal is expected.

To estimate the number of background events in the signal region from random combinations and sequential decays, the $m_{\mu\mu}$ distribution in the mass sidebands is extrapolated. The background from $B \to h^+h'^-$ decays is estimated separately. If the number of observed events in the signal region after unblinding it is consistent with the number of expected background events, a limit on $\mathscr{B}(B^0_{(s)} \to \mu^+\mu^-)$ is determined. The $B^+ \to J/\psi K^+$ decay with $J/\psi \to \mu^+\mu^-$ is taken as normalization channel because it has also two muons in the final state, has a high signal yield, and a known absolute branching ratio.

Several searches for $B^0_{(s)} \to \mu^+\mu^-$ decays were performed by both Tevatron experiments in Run II. In addition to the increased data sizes, the updated analyses also enhanced their sensitivity by improvements of the analysis technique. The left plot in Fig. 9.7 illustrates the evolution of the limit on the $B^0_s \to \mu^+\mu^-$ branching ratio for the various iterations of the analysis.

The first measurements by CDF and D0 were based on 171 pb^{-1} [20] and 240 pb^{-1} [21], respectively. The selection is optimized by simple requirements on a few variables. In this initial version of the analysis CDF uses the integrated luminosity measurement and the production cross section instead of the $B^+ \to J/\psi K^+$ mode for the normalization of the signal yield. Both experiments use simulation to determine the relative acceptances and efficiencies, CDF measures part of the relative efficiency on data. The trigger and muon identification efficiency is determined with $J/\psi \to \mu^+\mu^-$ events. The efficiency of the track reconstruction in the drift chamber is obtained from simulated tracks embedded in real data events. At the given levels of statistics, background from $B \to h^+h'^-$ decays is negligible and the distribution of combinatorial background is assumed to be flat. CDF uses background enriched control samples to verify the validity of the background estimation procedure. The control samples are like-sign dimuons, opposite-sign dimuons with negative decay length, and track pairs where at least one of the tracks fails the muon identification requirements. The inverted muon identification criterion decreases the signal efficiency by a factor 50 and increases the background yield by a factor three. Because CDF has a dimuon mass resolution of 24 MeV/c^2 it can resolve the B^0_s and

B^0 signal peaks which are 87 MeV/c^2 apart and measure $\mathscr{B}(B^0_s \to \mu^+\mu^-)$ and $\mathscr{B}(B^0 \to \mu^+\mu^-)$ individually. The analysis methods for B^0_s and B^0 are equivalent except that the signal mass windows are centered on the different nominal masses. The dimuon mass resolution of the D0 detector is about 120 MeV/c^2 so that the B^0_s and B^0 peaks overlap. Therefore D0 only quotes a limit on $\mathscr{B}(B^0_s \to \mu^+\mu^-)$ under the assumption that the yield of $B^0 \to \mu^+\mu^-$ events is negligible. This gives a conservative limit because if B^0 events would be present in data they would lead to a worse limit. Both experiments observe event yields in the signal regions that are consistent with background and set limits on $\mathscr{B}(B^0_{(s)} \to \mu^+\mu^-)$.

A more sophisticated selection is used in updated analyses on 364 pb^{-1} by CDF [22] and 1.3 fb^{-1} by D0 [23]. Up to five selection variables are combined in a likelihood ratio. In addition CDF increases the acceptance by including muon pairs where one of the muons is detected in the forward muon chamber. D0 explicitly accounts for sequential b-hadron decays in the background model. Again, consistency with the background only hypothesis is observed and limits are quoted.

Further improvements were introduced in analyses using 2 fb^{-1} (CDF) [24] and 7 fb^{-1} (D0) [25]. The muon identification of the CDF detector is enhanced by using information from the calorimeter, the matching of the track extrapolation to signals in the muon detector, and the ionization energy loss in the drift chamber combined in a likelihood function [26]. D0 and CDF each combine several selection variables in an artificial neural network. The advantage of the neural network is that correlations, which are difficult to model in a likelihood ratio approach, are no problem and can even be exploited to improve the separation between signal and background. CDF quotes an improvement of 25 % better background rejection at the same signal efficiency compared to the likelihood ratio used in the previous iteration of this analysis. CDF also checked that the network output distribution in data is well described by simulation with $B^+ \to J/\psi K^+$ events. A further increase in sensitivity, of 15 % in case of CDF, is achieved by splitting the signal region in bins of mass and network output. This exploits the fact that the signal to background ratio depends on both variables. The sensitivity has reached a level where $B \to h^+h'^-$ background has to be considered. It is estimated using efficiencies from simulation, misidentification probabilities measured with $D^0 \to K^-\pi^+$ events, and known branching ratios. The growing sensitivity of the analyses is illustrated by the fact that CDF expects about 0.5 and D0 about 2.7 $B^0_s \to \mu^+\mu^-$ signal events in their signal region if the branching ratio is as predicted in the standard model. Both experiment see no signal excess and set limits. The D0 result is $\mathscr{B}(B^0_s \to \mu^+\mu^-) < 5.1 \times 10^{-8}$ at 95 % confidence level [25]. The invariant mass and network output distributions are shown in Fig. 9.5.

The most recent $B^0_s \to \mu^+\mu^-$ search by CDF uses 7 fb^{-1} [27]. The signal to background discrimination power is again improved by combining 14 variables in a neural network with sophisticated preprocessing [28]. It is verified that the network does not bias the dimuon mass distribution and that it does not learn statistical fluctuations in the training data. The background extrapolation uses only events above a mass of 5 GeV/c^2 to exclude sequential decays. As in the previous iterations

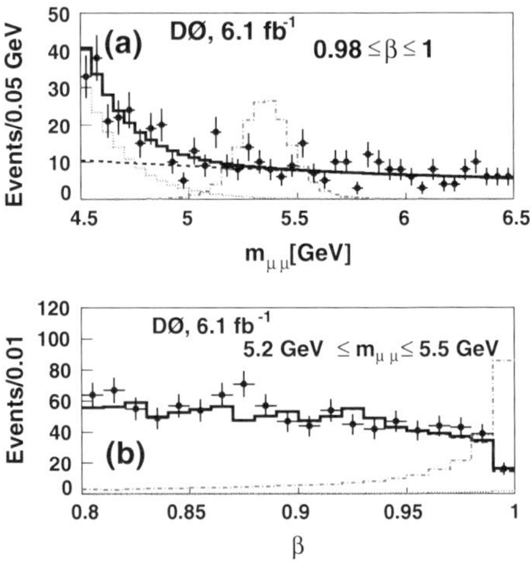

Fig. 9.5 Dimuon invariant mass (*top*) and network output distribution, β (*bottom*) of $B_s^0 \to \mu^+\mu^-$ candidates measured by D0 [25]. The *dash-dotted line* shows the shape of a hyothetical signal

of this analysis the background estimation procedure is checked with the background enriched control samples. The agreement of the data with the background hypothesis is quoted in terms of a p-value which is determined from the ratio of likelihoods between the signal plus background and background only hypothesis. The p-value for B^0 is 23 %, indicating good agreement with the background hypothesis, and a limit of $\mathscr{B}(B^0 \to \mu^+\mu^-) = 6.0 \times 10^{-9}$ at 95 % confidence level is obtained [27]. An excess of events is seen in the B_s^0 signal window as shown in Fig. 9.6. The probability that this is caused by a background fluctuation is 0.27 %. The low p-value is mainly driven by the events in the highest network output bin, $\nu_N > 0.995$. If only the two highest network bins are considered so that the bin at $0.970 < \nu_N < 0.987$ with a further excess is excluded the p-value is still low at 0.66 %. The excess in the third highest ν_N bin was carefully checked and could only be explained by a statistical fluctuation. Given the marginal agreement with the background hypothesis CDF quotes a point estimate of $\mathscr{B}(B_s^0 \to \mu^+\mu^-) = [1.8^{+1.1}_{-0.9}] \times 10^{-8}$, where the errors include the statistical and systematic uncertainties [27].

The $B_s^0 \to \mu^+\mu^-$ branching ratio measured by CDF is 5.6 times higher than the standard model prediction. While this sounds like a strong deviation from the standard model one has to take into account the uncertainties. The p-value of the standard model signal plus background hypothesis is 1.9 %. Thus there is no evidence for new physics. A reasonable explanation for the excess would be that an upward fluctuation of background and standard model signal is observed. Recent measurements at the LHC [29–31] seem to confirm this scenario. A comparison of $B_s^0 \to \mu^+\mu^-$ results is shown in the right plot of Fig. 9.7.

CDF also searched for the rare decays $B_{(s)}^0 \to e^+e^-$ and $B_{(s)}^0 \to e^\pm\mu^\mp$ in 2 fb^{-1} of data [32]. Because the electron is lighter than the muon the helicity suppression in the $B_{(s)}^0 \to e^+e^-$ decay is much stronger than in the $B_{(s)}^0 \to \mu^+\mu^-$ decay leading to a predicted branching ratio of the order of 10^{-15} [33, 34]. The decay $B_{(s)}^0 \to e^\pm\mu^\mp$ violates lepton family number conservation. So far such an effect

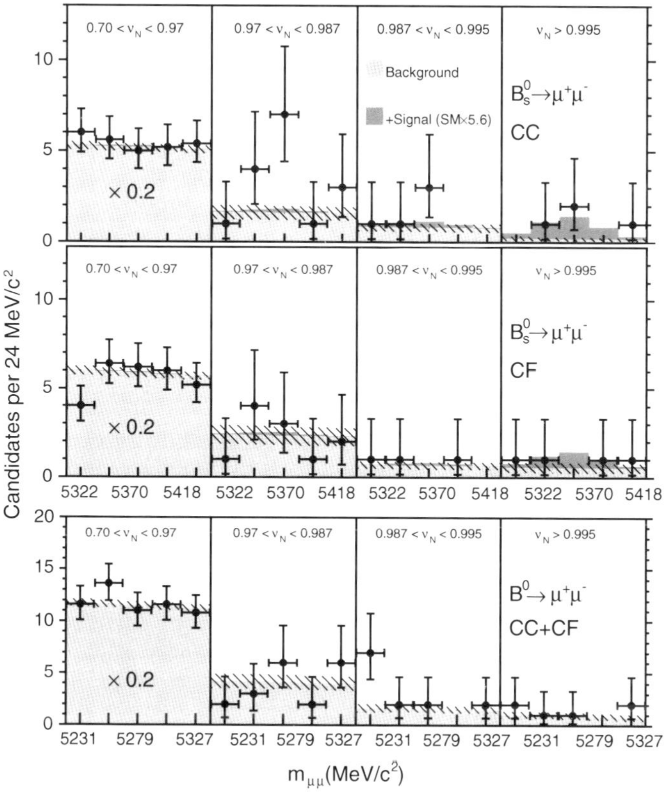

Fig. 9.6 Dimuon invariant mass distribution of $B^0_s \to \mu^+\mu^-$ (*top*) and $B^0 \to \mu^+\mu^-$ candidates (*bottom*) in bins of neural network output measured by CDF [27]. The $B^0_s \to \mu^+\mu^-$ data is shown separately for events with both muons in the central region of $|\eta| < 0.6$ (CC) and one muon in the central and one in the forward region of $0.6 < |\eta| < 1$ (CF)

was only observed in neutrino mixing [35, 36]. Branching ratios at the order of the experimental sensitivity are possible in new physics models like the one proposed by J. Pati and A. Salam [37]. In their theory the lepton number is treated as a fourth color and spin-1 gauge bosons, called leptoquarks, mediate transitions between leptons and quarks. If the coupling of the leptoquarks is not limited to quark-lepton pairs of the same generation, the $B^0_{(s)} \to e^\pm\mu^\mp$ decays can be described by the exchange of a single leptoquark as shown in Fig. 9.8. Different types of leptoquarks would mediate the B^0_s and B^0 decay.

The analysis technique is similar to the one used for $B^0_{(s)} \to \mu^+\mu^-$ decays. A peak in the dilepton invariant mass spectrum is searched for with the background prediction obtained from an extrapolation of event yields in mass sidebands and hadron misidentification rates applied to $B \to h^+h'^-$ decays. The analyses differ in the used trigger. While $B^0_{(s)} \to \mu^+\mu^-$ decays are selected by a dimuon trigger, $B^0_{(s)} \to e^+e^-$ and $B^0_{(s)} \to e^\pm\mu^\mp$ decays are collected by the Two-Track-Trigger.

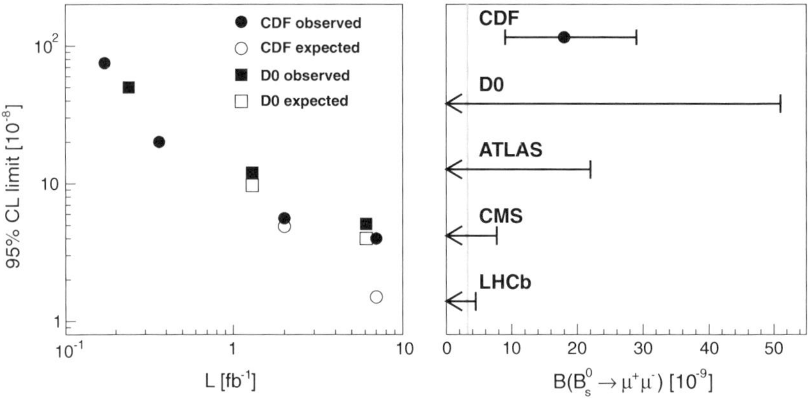

Fig. 9.7 Evolution of Tevatron limits on $\mathscr{B}(B_s^0 \to \mu^+\mu^-)$ (*left*) and comparison of measurements and 95 % C.L. limits by CDF [27], D0 [25], ATLAS [29], CMS [30], and LHCb [31] (*right*)

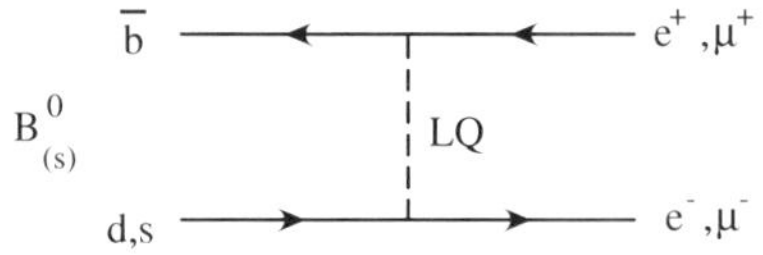

Fig. 9.8 Feynman diagram of a $B^0_{(s)} \to e^{\pm}\mu^{\mp}$ decay mediated by the exchange of a leptoquark (*LQ*)

Therefore the normalization mode in this analysis is $B^0 \to K^+\pi^-$. The electron identification is based on the ionization energy loss measurement in the drift chamber and the shower shape in the calorimeter.

As one can see in Fig. 9.9, the numbers of observed events in the B^0 and B_s^0 signal windows are consistent with the background expectations. The derived 90 % credibility level limits of $\mathscr{B}(B_s^0 \to e^+e^-) < 2.8 \times 10^{-7}$, $\mathscr{B}(B^0 \to e^+e^-) < 8.3 \times 10^{-8}$, $\mathscr{B}(B_s^0 \to e^+\mu^-) < 2.0 \times 10^{-7}$, and $\mathscr{B}(B^0 \to e^+\mu^-) < 6.4 \times 10^{-8}$ [32] are the most stringent to date. Limits on the leptoquark masses in the Pati-Salam model are calculated which are complementary to the limits obtained from direct searches [38, 39].

9.3 $B \to K^{(*)}\mu^+\mu^-$

The transition $b \to s\mu^+\mu^-$ is sensitive to new physics contributions because it is a flavor-changing neutral current and thus forbidden at tree level in the standard model. Feynman diagrams of the leading order standard model processes are depicted in Fig. 9.10. Experimentally the $b \to s\mu^+\mu^-$ transition is best studied in exclusive $B \to h\mu^+\mu^-$ decays, where B stands for B^+, B^0, B_s^0, or Λ_b^0 and h for K^+ or K^{*+}, K^{*0} or K_S^0, ϕ, or Λ, respectively. While the decays of the first two B meson types are also studied at the B-factories, the latter two became accessible only at the Tevatron.

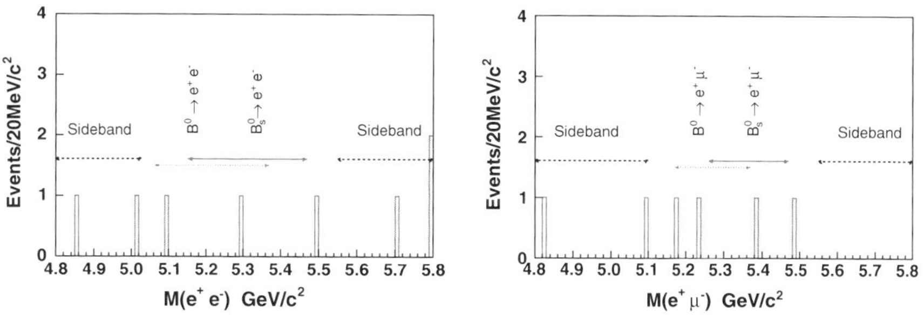

Fig. 9.9 Invariant mass distribution of $B^0_{(s)} \to e^+e^-$ (*left*) and $B^0_{(s)} \to e^{\pm}\mu^{\mp}$ candidates (*right*) [32]. The horizontal lines in the middle of the plots indicate the B^0 and B_s^0 signal windows

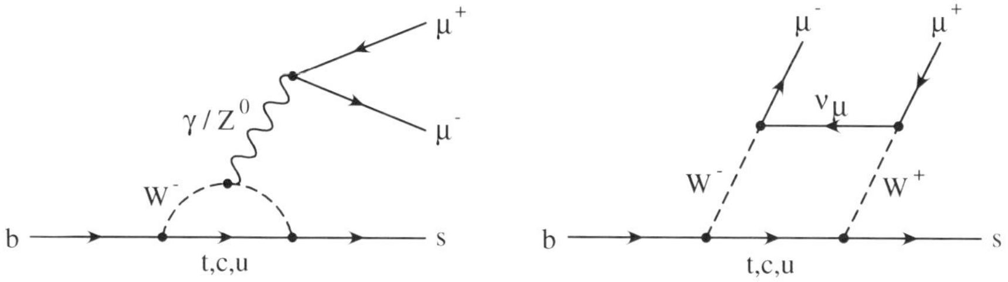

Fig. 9.10 standard model diagrams of $b \to s\mu^+\mu^-$ transitions

Several parameters can be measured in the $B \to h\mu^+\mu^-$ decays and allow to search for deviations from the standard model expectation or to constrain new physics models. The parameters usually depend on the squared dimuon invariant mass, $q^2 = m^2_{\mu^+\mu^-}$. An observable, that is provided by all decay modes, is the muon forward-backward asymmetry, A_{FB}. It can be determined from the $\cos\theta_\mu$ distribution, where θ_μ is the angle between the μ^+ (μ^-) momentum and the opposite B ($\bar{B}$) momentum in the dimuon rest frame. If the hadron h is a vector meson further kinematic observables become available, that allow to measure additional parameters. The longitudinal polarization of a K^* meson, F_L, determines the distribution of its helicity angle, θ_K, which is defined as the angle between kaon daughter particle and opposite B momentum direction in the K^* rest frame. The transverse polarization asymmetry, $A_T^{(2)}$, and the T-odd CP asymmetry, A_{im}, can be extracted from the distribution of the angle between the dimuon and K^* decay planes in the B rest frame, $\Delta\phi$. Definitions of the parameters and a derivation of the differential decay rates of

$$\frac{1}{\Gamma}\frac{d\Gamma}{d\cos\theta_K} = \frac{3}{2}F_L\cos^2\theta_K + \frac{3}{4}(1-F_L)(1-\cos^2\theta_K) \tag{9.7}$$

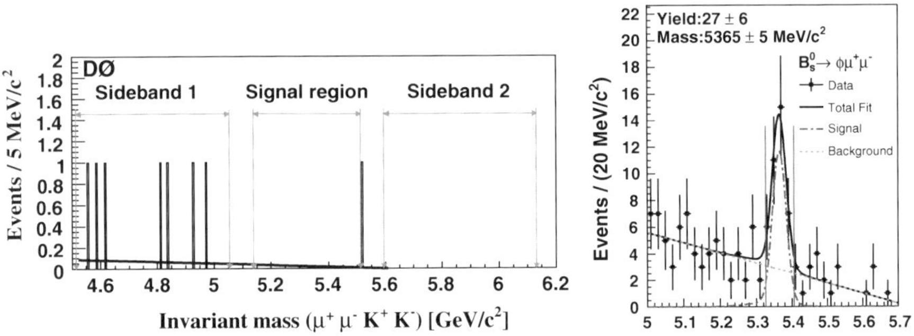

Fig. 9.11 Invariant mass distribution of $B_s^0 \to \phi\mu^+\mu^-$ candidates reconstructed by D0 with 450 pb^{-1} [41] (*left*) and by CDF with 4.4 fb^{-1} [42] (*right*)

$$\frac{1}{\Gamma}\frac{d\Gamma}{d\cos\theta_\mu} = \frac{3}{4}F_L(1-\cos^2\theta_\mu) + \frac{3}{8}(1-F_L)(1+\cos^2\theta_\mu) + A_{FB}\cos\theta_\mu \tag{9.8}$$

$$\frac{1}{\Gamma}\frac{d\Gamma}{d\Delta\phi} = \frac{1}{2\pi}\left[1 + \frac{1}{2}(1-F_L)A_T^{(2)}\cos 2\Delta\phi A_{im}\sin 2\Delta\phi\right] \tag{9.9}$$

can for example be found in Ref. [40].

The analyses of $B \to h\mu^+\mu^-$ decays discriminate signal from background based on variables like the decay length significance, the pointing angle, and the isolation. Signal yields are determined from peaks in the B candidate invariant mass. Since the resonant decays $B \to \psi h$, where ψ is J/ψ or ψ' and decays to two muons, have the same final state, but proceed mainly via tree-level diagrams they are rejected from the signal sample by excluding the corresponding q^2 regions. Because of the similarity to the signal modes and their high yield the decays $B \to J/\psi h$ are used as normalization in the calculation of a branching ratio. Relative efficiencies are determined from simulation.

The first search for the decay $B_s^0 \to \phi\mu^+\mu^-$ with $\phi \to K^+K^-$ at Run II was performed by D0 with a data sample corresponding to an integrated luminosity of 450 pb^{-1} [41]. The number of observed events in the signal region is consistent with the background expectation obtained from a linear extrapolation of sideband events (see Fig. 9.11). The derived 95 % confidence level limit on the relative branching ratio is $\mathscr{B}(B_s^0 \to \phi\mu^+\mu^-)/\mathscr{B}(B_s^0 \to J/\psi\phi) < 4.4 \times 10^{-3}$ [41].

A first hint of the $B_s^0 \to \phi\mu^+\mu^-$ decay was seen by CDF in an analysis of 924 pb^{-1} of data [43]. The analysis also considered the decays $B^+ \to K^+\mu^+\mu^-$ and $B^0 \to K^{*0}\mu^+\mu^-$ with $K^{*0} \to K^+\pi^-$. Additional vetos against specific backgrounds are applied. Decays of $B \to \psi h$ with $\psi \to \mu^+\mu^-\gamma$ are rejected by exploiting the fact that the reconstructed B and dimuon candidate masses are shifted due to the not reconstructed photon by a similar amount. Furthermore two- and three-track combinations that are consistent with a ψ, D^0, D^+, or D_s^+ decay are vetoed. The estimate of combinatorial background in the signal region is obtained from an extrapolation from the upper mass sideband using a shape derived from a background enriched sample. Simulation is used to estimate the background from hadrons misidentified as muons and from cross feed between $B_s^0 \to \phi\mu^+\mu^-$ and $B^0 \to K^{*0}\mu^+\mu^-$ caused by the wrong mass assignment of a kaon or pion track. Signals with a significance of 4.5, 2.9, and 2.4 standard deviations are observed for B^+, B^0, and B_s^0, respectively. Relative and absolute branching ratio measurements are quoted and a limit of $\mathscr{B}(B_s^0 \to \phi\mu^+\mu^-)/\mathscr{B}(B_s^0 \to J/\psi\phi) < 2.6 \times 10^{-3}$ at 95 % confidence level is reported [43].

With a data sample of 4.4 fb^{-1} CDF succeeded to observe the $B_s^0 \to \phi\mu^+\mu^-$ decay for the first time with a significance of 6.3σ [42] (see Fig. 9.11). In this analysis the muon identification is improved by the likelihood approach [26] used also in the $B_{(s)}^0 \to \mu^+\mu^-$ analysis which reduces the background from misidentified hadrons to a negligible level. The discrimination of signal and background is further enhanced by combining several selection variables in an artificial neural network. Signal yields are obtained from a fit to the B candidate invariant mass distribution. The measured relative B_s^0 branching ratio is $\mathscr{B}(B_s^0 \to \phi\mu^+\mu^-)/\mathscr{B}(B_s^0 \to J/\psi\phi) = [1.11 \pm 0.25\ (\text{stat}) \pm 0.09\ (\text{syst})] \times 10^{-3}$ [42].

Total and differential branching ratios in bins of q^2 are determined for B^+ and B^0 mesons. F_L is measured as a function of q^2 for the $B^0 \to K^{*0}\mu^+\mu^-$ decay in a fit to the $\cos\theta_K$ distribution. Using the measured F_L values, $A_{FB}(q^2)$ is determined from a fit to the $\cos\theta_\mu$ distribution. Events with a swapped kaon and pion assignment are accounted for by a component in the fit. The muon forward-backward asymmetry in $B^+ \to K^+\mu^+\mu^-$ is determined from a fit to the $\cos\theta_\mu$ distribution using Eq. (9.8) with $F_L = 1$ as signal model. An angle dependent efficiency is obtained from simulation and taken into account in the fits.

An increase of the integrated luminosity to 6.8 fb^{-1}, the addition of the decay $B^+ \to K^{*+}\mu^+\mu^-$ with $K^{*+} \to K_S^0\pi^+$ and $K_S^0 \to \pi^+\pi^-$, and a further optimization of the selection with a neural network made it possible to also measure $A_T^{(2)}$ and A_{im} for B^+ and B^0 mesons for the first time [44]. Both parameters are determined in bins of q^2 in a fit of the $\Delta\phi$ distribution. The results are presented in Fig. 9.12. All measurements agree well with the standard model expectation. First measurements at the B-factory saw a tendency towards positive A_{FB} values in the low q^2 region for $B \to K^*\mu^+\mu^-$ decays which favored new physics models, where the sign of the C_7 Wilson coefficient is inverted. This trend is not confirmed by the CDF measurement.

Using the same data sample and analysis technique CDF observed the rare baryonic decays $\Lambda_b^0 \to \Lambda\mu^+\mu^-$ with $\Lambda \to p\pi^-$ for the first time [46]. Figure 9.13 shows the Λ_b^0 signal which has a significance of 5.8σ. A further B^0 decay channel, $B^0 \to K_S^0\mu^+\mu^-$ with $K_S^0 \to \pi^+\pi^-$ is added as well. Cross feed between the two new modes is suppressed to a level below 1 % by a requirement on the asymmetry between the momentum component of the daughter particles in the direction of the mother particle momentum. Total and differential branching ratios are measured for the decays $\Lambda_b^0 \to \Lambda\mu^+\mu^-$, $B_s^0 \to \phi\mu^+\mu^-$, $B^+ \to K^+\mu^+\mu^-$, $B^0 \to K^{*0}\mu^+\mu^-$,

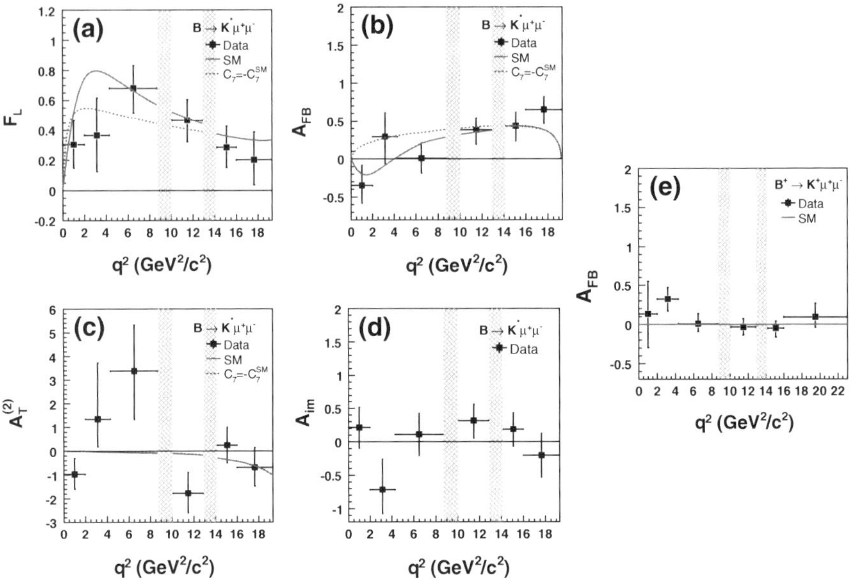

Fig. 9.12 Measurements of F_L (**a**), A_{FB} (**b**), $A_T^{(2)}$ (**c**), and A_{im} (**d**) for $B^0 \to K^{*0}\mu^+\mu^-$ and $B^+ \to K^{*+}\mu^+\mu^-$ decays combined and A_{FB} measurement for $B^+ \to K^+\mu^+\mu^-$ (**e**) [44]. The *solid curve* shows the standard model prediction [45] and the *dashed curve* the prediction on case of a flipped sign of the C_7 Wilson coefficient. The *vertical bands* indicate the excluded J/ψ and ψ' regions

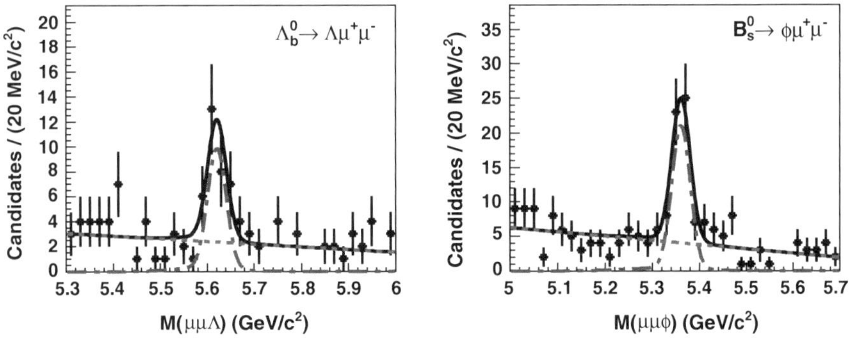

Fig. 9.13 Invariant mass distribution of $\Lambda_b^0 \to \Lambda\mu^+\mu^-$ (*left*) and $B_s^0 \to \phi\mu^+\mu^-$ candidates (*right*) measured by CDF [46]

$B^0 \to K^0\mu^+\mu^-$, and $B^+ \to K^{*+}\mu^+\mu^-$. Dominant systematic uncertainties come from the trigger efficiency and the unknown $\Lambda_b^0 \to J/\psi\,\Lambda$ polarization. In general good agreement with the standard model prediction is observed, only the Λ_b^0 branching ratio of $\mathscr{B}(\Lambda_b^0 \to \Lambda\mu^+\mu^-) = [1.73 \pm 0.42\,(\text{stat}) \pm 0.55\,(\text{syst})] \times 10^{-6}$ [46] is somewhat below the expectation of $[4.6 \pm 1.6] \times 10^{-6}$ [47].

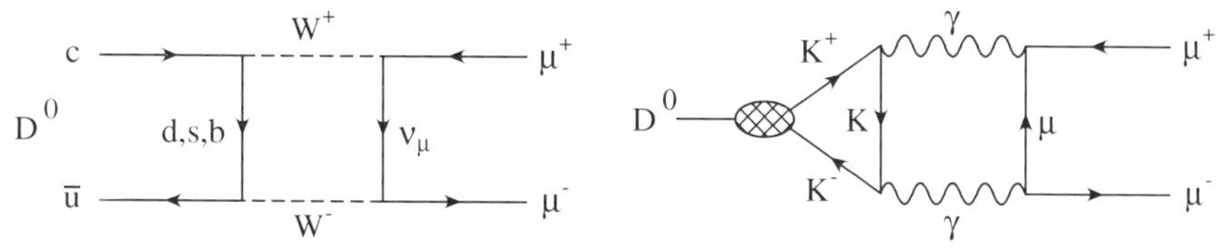

Fig. 9.14 Short distance (*left*) and long distance (*right*) diagrams of the $D^0 \to \mu^+\mu^-$ decay in the standard model

9.4 Charm Decays

Rare decays of charm hadrons provide complementary information to rare B decays because they are sensitive to new physics in the up quark sector. A promising decay channel is $D^0 \to \mu^+\mu^-$. The same Feynman diagrams as in $B^0_{(s)} \to \mu^+\mu^-$ decays contribute to $D^0 \to \mu^+\mu^-$ decays, except that down- and up-type quarks are exchanged. The left part of Fig. 9.14 shows a box diagram that corresponds to the $B^0_s \to \mu^+\mu^-$ box diagram in Fig. 9.4. Because the GIM mechanism [48] leads to a stronger suppression than for B meson decays where the heavy top quark counters the GIM cancellation, the short distance box and penguin processes provide only a small contribution to the total decay rate in the standard model. The dominant contributions come from long distance processes with intermediate hadronic states, like the process shown in the right part of Fig. 9.14. While the short distance processes could only account for a branching ratio of the order of 10^{-18}, the long distance processes can increase the branching ratio to about 4×10^{-13} [49]. However, this is still many orders of magnitude below the current experimental sensitivity. New physics contributions, in particular in the model of R-parity violating supersymmetry, where transitions between standard model and supersymmetric particles are allowed, could enhance the branching ratio up to this level [49].

The first search for $D^0 \to \mu^+\mu^-$ at the Tevatron was performed by CDF with 69 pb^{-1} of data [50]. The data is collected by the Two-Track-Trigger so that the kinematically very similar $D^0 \to \pi^+\pi^-$ decay can be used as normalization mode. The muons from the $D^0 \to \mu^+\mu^-$ candidates are identified by signals in the central muon detector. A strong background suppression is achieved by requiring the D^0 to come from a $D^{*+} \to D^0\pi^+$ decay. Selection requirements are optimized on a statistically independent sample. The relative efficiency and acceptance between signal and normalization mode are determined with simulation except for the muon identification efficiency which is measured with $J/\psi \to \mu^+\mu^-$ decays.

Like in the case of the $B^0_{(s)} \to \mu^+\mu^-$ analyses the signal is searched for by comparing the event yield in a dimuon invariant mass signal window, which is blinded until the analysis is finalized, to an estimation of background events. The background from $D^0 \to \pi^+\pi^-$ decays where both pions are misidentified as muons is determined from $D^0 \to \pi^+\pi^-$ events falling into the signal window if reconstructed

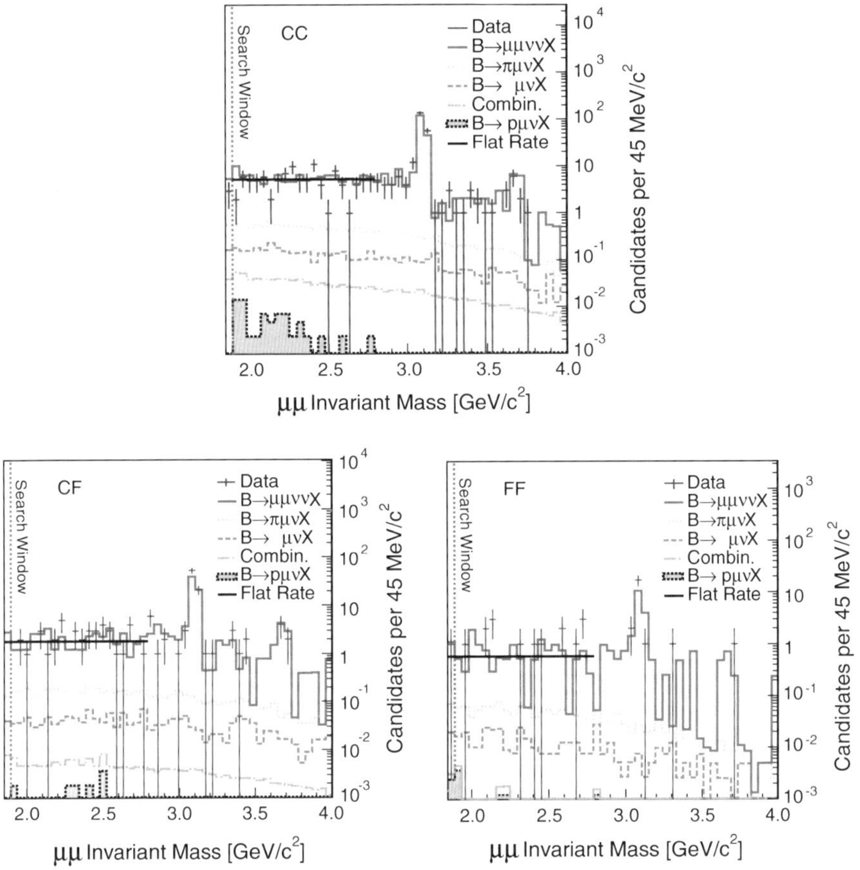

Fig. 9.15 Dimuon invariant mass distribution of $D^0 \to \mu^+\mu^-$ candidates measured by CDF in three different dimuon acceptance regions, where C stands for muons with $|\eta| < 0.6$ and F for muons with $0.6 < |\eta| < 1$, together with the background estimates [51]. The left most bin in each plot is the signal search window

with dimuon hypothesis and the pion misidentification probability measured with $D^0 \to K^-\pi^+$ events. Misidentified $D^0 \to K^-\pi^+$ or $D^0 \to K^+K^-$ events are shifted to lower masses so that they do not contribute to the events in the signal window. All other background is assumed to be flat in mass and extrapolated from the upper mass sideband. As the number of observed events in the signal window is consistent with the background expectation a limit of $\mathscr{B}(D^0 \to \mu^+\mu^-) < 3.3 \times 10^{-6}$ at 95 % confidence level is set [50].

An improved analysis method was applied by CDF on a data sample of 360 pb^{-1} [51]. The muon candidate acceptance is enlarged to $|\eta| < 1$ by including tracks in the central muon extension detector. The identification of muons is improved by the

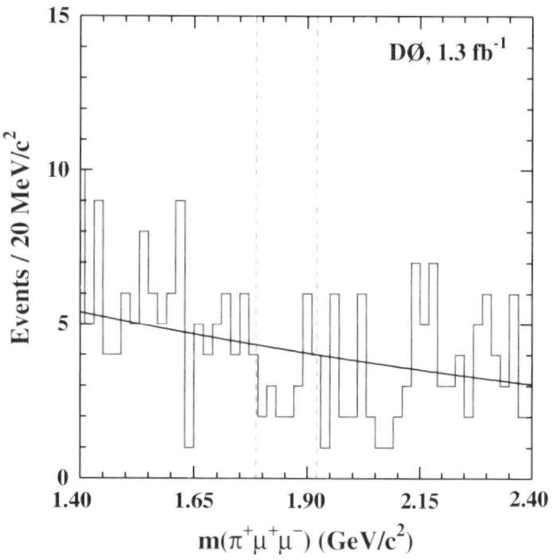

Fig. 9.16 Invariant mass spectrum of $D^+ \to \pi^+\mu^+\mu^-$ candidates measured by D0 [52]

muon likelihood function [26] that is also employed in the $B_s^0 \to \mu^+\mu^-$ search. As this reduces the background from misidentified hadrons the dominant background are now sequential semi-muonic B hadron decays, $B \to \mu^+\mu^- X$, with two real muons. Two variables are used to discriminate this kind of background from signal for which the majority of D^0 mesons is coming from the primary vertex. The D^0 impact parameter with respect to the primary vertex and the transverse decay length significance tend to be larger for $B \to \mu^+\mu^- X$ events than for prompt D^0 mesons. A requirement on a probability ratio constructed from both variables reduces the background by a factor four while keeping 87 % of the signal.

The amount of background in the signal window from $B \to \mu^+\mu^- X$ events and from semi-muonic B decays with a misidentified hadron is estimated from a sample of simulated B decays normalized to data and using measured muon identification efficiencies and misidentification probabilities. The misidentification probabilities are also used to estimate the combinatorial background from a sample of hadron pairs. Figure 9.15 shows the estimation of the background in the signal window and the upper sideband compared with data. Since the background above the signal window is consistent with a flat distribution, an extrapolation of sidebands events in data with a constant function is used as alternative method to estimate the background in the signal region and yields a consistent result. No excess in the signal mass region is seen in data and a limit of $\mathcal{B}(D^0 \to \mu^+\mu^-) < 3.0 \times 10^{-7}$ at 95 % confidence level is obtained [51].

An analogon to the $b \to s\mu^+\mu^-$ transition decays $B \to h\mu^+\mu^-$ discussed in Sect. 9.3 is the decay $D^+ \to \pi^+\mu^+\mu^-$. Also here the strong GIM suppression of the short distance flavor-changing neutral current transition $c \to u\mu^+\mu^-$ causes long distance processes to dominate the total decay rate. In the standard model a branching ratio up to the order 10^{-6} is predicted [49].

D0 searched for this decay in 1.3 fb^{-1} of data [52]. As a cross-check the branching ratio of the decay via an intermediate ϕ resonance, $D^+ \to \phi\pi^+ \to \pi^+\mu^+\mu^-$, is measured first. A signal with a significance of 4.1σ is observed in the $\phi\pi^+$ invariant mass distribution and a branching ratio that is consistent with the product of $\mathscr{B}(D^+ \to \phi\pi^+)$ and $\mathscr{B}(\phi \to \mu^+\mu^-)$ is obtained.

After excluding the ϕ mass region from the selection of dimuon pairs and optimizing the selection requirements on kinematic and vertex quality variables, the $\pi^+\mu^+\mu^-$ invariant mass distribution is inspected (see Fig. 9.16). The number of events observed in the signal region around the D^+ mass is consistent with the background estimation obtained from an extrapolation of the data in the sidebands. By normalizing to the resonant $D^+ \to \phi\pi^+ \to \pi^+\mu^+\mu^-$ decay a limit on the branching ratio of $\mathscr{B}(D^+ \to \pi^+\mu^+\mu^-) < 3.9 \times 10^{-6}$ at 90 % confidence level is derived [52].

References

1. T. Aaltonen et al., Phys. Rev. Lett. **108**, 211803 (2012)
2. A. Abulencia et al., Phys. Rev. Lett. **97**, 211802 (2006)
3. E. Barberio, Z. Was, Comput. Phys. Commun. **79**, 291 (1994)
4. T. Aaltonen et al., Phys. Rev. Lett. **106**, 181802 (2011)
5. M. Beneke, M. Neubert, Nucl. Phys. **B675**, 333 (2003)
6. A. Ali, G. Kramer, Y. Li, C.D. Lu, Y.L. Shen et al., Phys. Rev. **D76**, 074018 (2007)
7. A.R. Williamson, J. Zupan, Phys. Rev. **D74**, 014003 (2006)
8. J.f. Sun, G.h. Zhu, D.s. Du, Phys. Rev. D68, 054003 (2003).
9. A.J. Buras, R. Fleischer, S. Recksiegel, F. Schwab, Nucl. Phys. **B697**, 133 (2004)
10. T. Aaltonen et al., Phys. Rev. Lett. **103**, 031801 (2009)
11. D.E. Acosta et al., Phys. Rev. **D72**, 051104 (2005)
12. E. Gamiz, C.T. Davies, G.P. Lepage, J. Shigemitsu, M. Wingate, Phys. Rev. **D80**, 014503 (2009)
13. A.J. Buras, M.V. Carlucci, S. Gori, G. Isidori, JHEP **1010**, 009 (2010)
14. K. de Bruyn, R. Fleischer, R. Knegjens, P. Koppenburg, M. Merk, et al., (2012).
15. S.P. Martin, (1997). In book 'Perspectives on Supersymmetry II', G. Kane, ed., World Scientific.
16. G.G. Ross, L. Velasco-Sevilla, O. Vives, Nucl. Phys. **B692**, 50 (2004)
17. K. Agashe, C.D. Carone, Phys. Rev. **D68**, 035017 (2003)
18. A. Dedes, H.K. Dreiner, U. Nierste, Phys. Rev. Lett. **87**, 251804 (2001)
19. A.J. Buras, Acta Phys. Polon. **B41**, 2487 (2010)
20. D.E. Acosta et al., Phys. Rev. Lett. **93**, 032001 (2004)
21. V.M. Abazov et al., Phys. Rev. Lett. **94**, 071802 (2005)
22. A. Abulencia et al., Phys. Rev. Lett. **95**, 221805 (2005)
23. V.M. Abazov et al., Phys. Rev. **D76**, 092001 (2007)
24. T. Aaltonen et al., Phys. Rev. Lett. **100**, 101802 (2008)
25. V.M. Abazov et al., Phys. Lett. **B693**, 539 (2010)
26. G.A. Giurgiu, FERMILAB-THESIS-2005-41, UMI-31-86027 (2005).
27. T. Aaltonen et al., Phys. Rev. Lett. **107**, 239903 (2011)
28. M. Feindt, U. Kerzel, Nucl. Instrum. Meth. **A559**, 190 (2006)
29. G. Aad et al., Phys. Lett. **B713**, 387 (2012)
30. S. Chatrchyan et al., JHEP **1204**, 033 (2012)
31. R. Aaij et al., Phys. Rev. Lett. **108**, 231801 (2012)
32. T. Aaltonen et al., Phys. Rev. Lett. **102**, 201801 (2009)
33. M. Misiak, J. Urban, Phys. Lett. B**451**, 161 (1999)
34. G. Buchalla, A.J. Buras, Nucl. Phys. **B548**, 309 (1999)

35. Y. Fukuda et al., Phys. Rev. Lett. **81**, 1562 (1998)
36. Q. Ahmad et al., Phys. Rev. Lett. **87**, 071301 (2001)
37. J.C. Pati, A. Salam, Phys. Rev. **D10**, 275 (1974)
38. A. Abulencia et al., Phys. Rev. **D73**, 051102 (2006)
39. V. Abazov et al., Phys. Lett. **B671**, 224 (2009)
40. U. Egede, T. Hurth, J. Matias, M. Ramon, W. Reece, JHEP **0811**, 032 (2008)
41. V.M. Abazov et al., Phys. Rev. **D74**, 031107 (2006)
42. T. Aaltonen et al., Phys. Rev. Lett. **106**, 161801 (2011)
43. T. Aaltonen et al., Phys. Rev. **D79**, 011104 (2009)
44. T. Aaltonen et al., Phys. Rev. Lett. **108**, 081807 (2012)
45. C. Bobeth, G. Hiller, D. van Dyk, C. Wacker, JHEP **1201**, 107 (2012)
46. T. Aaltonen et al., Phys. Rev. Lett. **107**, 201802 (2011)
47. T. Aliev, K. Azizi, M. Savci, Phys. Rev. **D81**, 056006 (2010)
48. S.L. Glashow, J. Iliopoulos, L. Maiani, Phys. Rev. **D2**, 1285 (1970)
49. G. Burdman, E. Golowich, J.L. Hewett, S. Pakvasa, Phys. Rev. **D66**, 014009 (2002)
50. D.E. Acosta et al., Phys. Rev. **D68**, 091101 (2003)
51. T. Aaltonen et al., Phys. Rev. **D82**, 091105 (2010)
52. V.M. Abazov et al., Phys. Rev. Lett. **100**, 101801 (2008)

Chapter 10
Conclusions

A new era of flavor physics was entered at the Run II of the Tevatron $p\bar{p}$ collider, precision measurements of B_s^0 meson properties. A highlight is certainly the direct observation of B_s^0 oscillations and the measurement of the oscillation frequency Δm_s in the year 2006 [1]. While the B_s^0 system, consisting of quarks of the second and third generation, was initially considered a place where sizable new physics effects could be present, the Δm_s result agrees well with the standard model expectation. As large new physics contributions to the absolute size of the B_s^0 mixing amplitude are therefore excluded, the next task was to measure the phase of the mixing amplitude where large deviations from the standard model value were still possible. First measurements seemed indeed to point into this direction [2, 3]. However the most recent results by CDF [4], D0 [5], and LHCb [6] show better agreement with the standard model and suggest that the initial discrepancy at a level of 2σ was a statistical fluctuation. Several further measurements complement the determination of the *CP* violating phase via a time-dependent angular analysis of $B_s^0 \to J/\psi\phi$ decays. These include B_s^0 lifetime measurements in decays to flavor specific and *CP* eigenstates, like $J/\psi f_0(980)$ [7], and branching ratio measurements of $B_s^0 \to D_s^{(*)+}D_s^{(*)-}$ [8, 9] which are sensitive to the decay width difference $\Delta\Gamma$.

A further high impact measurement at the Tevatron is the search for the rare decay $B_s^0 \to \mu^+\mu^-$ [10, 11] where large enhancements compared to the standard model branching ratio were predicted by several new physics models. During Run II the limits on $\mathscr{B}(B_s^0 \to \mu^+\mu^-)$ were improved by more than an order of magnitude, reaching almost the sensitivity of the standard model branching ratio. New physics effects that would increase the branching ratio by a large factor are excluded. A first hint of a signal, consistent with coming from a standard model process, is seen by CDF [11]. Signals of other rare decays, like $B_s^0 \to K^+K^-$ [12] and $B_s^0 \to \phi\mu^+\mu^-$ [13], could be established for the first time and their branching ratios are measured.

The Tevatron did not only advance flavor physics in the B_s^0 meson sector, but also extended it on the field of heavier hadrons. The lifetime of the B_c^+ [14, 15], the only meson with two distinct heavy quarks, and the Λ_b^0 baryon [16–19] are measured. The

T. Kuhr, *Flavor Physics at the Tevatron*, Springer Tracts in Modern Physics 249,
DOI: 10.1007/978-3-642-10300-1_10, © Springer-Verlag Berlin Heidelberg 2013

Λ_b^0 lifetime puzzle of measured lifetimes below the predictions seems to be solved by recent theoretical calculations, but the CDF measurement in the decay mode $\Lambda_b^0 \to J/\psi \Lambda$ favors a value consistent with the original predictions. The large Λ_b^0 production rate at the Tevatron also made it possible to observe charmless and $b \to s\mu^+\mu^-$ transition decays of Λ_b^0 baryons for the first time [20, 21] and to measure *CP* asymmetries of the former ones [22]. These measurements of heavy baryon properties complement the information from the corresponding rare B meson decays.

The properties of B^0 and B^+ mesons are well studied by the B-factory experiments BaBar and Belle. But also on this sector the Tevatron experiments could provide results that are competitive, like lifetime, branching ratio, and *CP* asymmetry measurements. Some parameters, like the T-odd *CP* asymmetry in $B \to K^*\mu^+\mu^-$ decays, were even measured first at CDF [23]. In the search for $B^0 \to \mu^+\mu^-$ CDF played the leading role [11] until the LHC experiments took over.

Finally, important contributions on the charm sector are provided by the Tevatron, like the evidence for D^0 oscillations [24] and the limit on $D^0 \to \mu^+\mu^-$ decays [25]. The CDF measurement of the *CP* asymmetries in $D^0 \to K^+K^-$ and $D^0 \to \pi^+\pi^-$ decays [26] confirms the unexpectedly large difference between these two values measured by LHCb. Whether this is consistent with the standard model or a hint for new physics is being discussed among theorists.

In general most measurements by CDF and D0 agree well with the predictions of the standard model. Therefore the Tevatron results have considerably tightened the parameter space of several new physics models, as for example discussed in Ref. [27]. The most prominent remaining discrepancy with the standard model expectation is the like-sign dimuon charge asymmetry measured by D0 [28]. Theoretical models have difficulties to explain the large value by new physics in the B_s^0 system because it would require an increase of $\Delta\Gamma$, that is inconsistent with other measurements. An independent confirmation of the experimental result is therefore essential.

This confirmation could come from LHCb to which the baton was passed on. Many LHCb measurements will be based upon the experimental techniques developed at the Tevatron for flavor physics measurements at hadron colliders, like flavor tagging algorithms. A key feature, that determines the kind of measurements that can be performed, is the trigger system. For example B_s^0 oscillations could only be observed by CDF because of its displaced track trigger which collected events with hadronic B_s^0 decays.

The competition between CDF and D0, and between the Tevatron and B-factories boosted the output of high profile physics results and provided independent cross checks of various results. In the next years, experimental flavor physics will be dominated by LHCb. If LHCb finds evidence for new physics an independent cross check may only be possible when the super flavor factory experiments SuperB in Italy and Belle II in Japan have started to take data. The next generation B factory experiments are also needed to cover areas, like B decays involving neutrinos, which are not accessible at hadron colliders. Together the hadron and lepton collider experiments will yield a broad spectrum of flavor physics results, that will help us to obtain a better understanding of the fundamental particles and interactions in nature.

References

1. A. Abulencia et al., Phys. Rev. Lett. **97**, 242003 (2006)
2. T. Aaltonen et al., Phys. Rev. Lett. **100**, 161802 (2008)
3. V.M. Abazov et al., Phys. Rev. Lett. **101**, 241801 (2008)
4. T. Aaltonen et al., Phys. Rev. **D85**, 072002 (2012)
5. V.M. Abazov et al., Phys. Rev. **D85**, 032006 (2012)
6. R. Aaij et al., Phys. Rev. Lett. **108**, 101803 (2012)
7. T. Aaltonen et al., Phys. Rev. **D84**, 052012 (2011)
8. V.M. Abazov et al., Phys. Rev. Lett. **102**, 091801 (2009)
9. T. Aaltonen et al., Phys. Rev. Lett. **108**, 201801 (2012)
10. V.M. Abazov et al., Phys. Lett. **B693**, 539 (2010)
11. T. Aaltonen et al., Phys. Rev. Lett. **107**, 239903 (2011)
12. A. Abulencia et al., Phys. Rev. Lett. **97**, 211802 (2006)
13. T. Aaltonen et al., Phys. Rev. Lett. **106**, 161801 (2011)
14. A. Abulencia et al., Phys. Rev. Lett. **97**, 012002 (2006)
15. V.M. Abazov et al., Phys. Rev. Lett. **102**, 092001 (2009)
16. V.M. Abazov et al., Phys. Rev. Lett. **99**, 182001 (2007)
17. V.M. Abazov et al., Phys. Rev. Lett. **99**, 142001 (2007)
18. T. Aaltonen et al., Phys. Rev. Lett. **106**, 121804 (2011)
19. T. Aaltonen et al., Phys. Rev. Lett. **104**, 102002 (2010)
20. T. Aaltonen et al., Phys. Rev. Lett. **103**, 031801 (2009)
21. T. Aaltonen et al., Phys. Rev. Lett. **107**, 201802 (2011)
22. T. Aaltonen et al., Phys. Rev. Lett. **106**, 181802 (2011)
23. T. Aaltonen et al., Phys. Rev. Lett. **108**, 081807 (2012)
24. T. Aaltonen et al., Phys. Rev. Lett. **100**, 121802 (2008)
25. T. Aaltonen et al., Phys. Rev. **D82**, 091105 (2010)
26. T. Aaltonen et al., Phys. Rev. **D85**, 012009 (2012)
27. W. Altmannshofer, A.J. Buras, S. Gori, P. Paradisi, D.M. Straub, Nucl. Phys. **B830**, 17 (2010)
28. V.M. Abazov et al., Phys. Rev. **D84**, 052007 (2011)

Printed by Printforce, the Netherlands